Developments in Economic Geology 12

GEOTHERMAL RESOURCES: AN ENERGY ALTERNATIVE

Further titles in this series

1. *I.L. ELLIOTT and W.K. FLETCHER (Editors)*
GEOCHEMICAL EXPLORATION 1974

2. *P.M.D. BRADSHAW (Editor)*
CONCEPTUAL MODELS IN EXPLORATION GEOCHEMISTRY
The Canadian Cordillera and Canadian Shield

3. *G.J.S. GOVETT and M.H. GOVETT (Editors)*
WORLD MINERAL SUPPLIES
Assessment and Perspective

4. *R.T. SHUEY*
SEMICONDUCTING ORE MINERALS

5. *J.S. SUMNER*
PRINCIPLES OF INDUCED POLARIZATION FOR GEOPHYSICAL EXPLORATION

6. *R.A. RICH, H.D. HOLLAND and U. PETERSEN*
HYDROTHERMAL URANIUM DEPOSITS

7. *J.G. MORSE (Editor)*
NUCLEAR METHODS IN MINERAL EXPLORATION AND PRODUCTION

8. *M. KUZVART and M. BÖHMER*
PROSPECTING AND EXPLORATION FOR MINERAL DEPOSITS

9. *C.R.M. BUTT and I.G.P. WILDING (Editors)*
GEOCHEMICAL EXPLORATION 1976

10. *G.B. FETTWEIS*
WORLD COAL RESOURCES
Methods of Assessment and Results

11. *R.G. TAYLOR*
GEOLOGY OF TIN DEPOSITS

Developments in Economic Geology 12

geothermal resources: an energy alternative

by

Harsh K. Gupta

Assistant Director
National Geophysical Research Institute, Hyderabad, India
and
Adjunct Professor, Institute for Geosciences,
University of Texas at Dallas, Texas, U.S.A.

ELSEVIER SCIENTIFIC PUBLISHING COMPANY
Amsterdam — Oxford — New York 1980

ELSEVIER SCIENTIFIC PUBLISHING COMPANY
335 Jan van Galenstraat
P.O. Box 211, 1000 AE Amsterdam, The Netherlands

Distributors for the United States and Canada:

ELSEVIER/NORTH-HOLLAND INC.
52, Vanderbilt Avenue
New York, N.Y. 10017

Library of Congress Cataloging in Publication Data

Gupta, Harsh K 1942-
Geothermal resources.

(Developments in economic geology ; 12)
Bibliography: p.
Includes index.
1. Geothermal engineering. 2. Geothermal resources. I. Title. II. Series.
TJ280.7.G86 621.44 80-14659
ISBN 0-444-41865-2

ISBN 0-444-41865-2 (Vol. 12)
ISBN 0-444-41250-6 (Series)

Printed in The Netherlands

To My Parents

Smt. Sushila Devi and
Shri Vishnu Kumar

FOREWORD

The monograph *Geothermal Resources: An Energy Alternative* by Harsh K. Gupta presents for the first time a readable, coherent account of all facets of geothermal energy development. The development of alternative, renewable energy resources during the past decade has been spurred by a crunch in the supplies of traditional hydrocarbon energy resources, which have previously been inexpensive and readily available. The result has been a hurried pace of research and development to bring alternative energy resources on line. Geothermal energy promises to play a role in supplying electric power, industrial process heating, and heat for buildings.

The book introduces the underlying concepts that are necessary for understanding the factors controlling the occurrence and characteristics of geothermal systems. Earth's structure and the distribution of radiogenic sources are integrated with the plate tectonic picture of the earth. The locations of volcanic activity both present and recent geologic past are presented in this plate tectonic context. The mechanisms of heat transfer are dominant factors in understanding the potential pitfalls in the interpretation of heat flow surveys and problems in producing geothermal energy.

Geothermal energy development draws together scientists from diverse traditional disciplines, such as geologists, mechanical engineers, geophysicists, and investors. While they may easily keep abreast of the developments in their specialty, they must be able to converse with those from different backgrounds. The account of these various fields is complete. The traditional surveys such as the seismic reflection technique are used less frequently in geothermal work than in hydrocarbon exploration, whereas seismic refraction or magnetotelluric are used considerably more. The technology related to drilling and reservoir modelling at high temperatures in hostile environments raises problems that are unique to the geothermal industry. The treatment and choice of material from each discipline reflects this emphasis.

After treating each of these disciplines in detail a unified case study is presented of the Cerro Prieto geothermal field in Mexico. From the exploration architecture used to establish a model of the geothermal system, Dr. Gupta leads us through the problems of drilling and producing this resource.

The up to date accounts of geothermal resource models; geological, geochemical, and geophysical exploration techniques; and drilling and production technology appear in diverse technical journals and proceedings of symposia. Dr. Gupta has extracted the important concepts and current technical developments from these sources and meshed them into a single volume for these individuals whether graduate students or scientists, engineers, or investors seeking a current account of this new technological field.

May, 1980

RONALD W. WARD
Director, Center for Energy Studies
University of Texas at Dallas

PREFACE

Although hot springs have been used for centuries for balneological purposes, the use of earth's heat as a source of energy began early in this century when electricity was generated for the first time from geothermal steam at Larderello, Italy, in 1904. The spread of technology to other parts of the world had been rather slow during the first half of this century. As a consequence of the continuously increasing demand for energy and the rapid depletion of other easily available sources of energy, geothermal resource development has received a tremendous impetus during the last two decades. The United Nations has contributed significantly to this development by periodically organizing conferences. These include the conference on New Sources of Energy held in Rome in 1961, the U.N. Symposium on the Development and Utilization of Geothermal Resources at Pisa, Italy in 1970, and the Second U.N. Symposium on the Development and Use of Geothermal Resources held at San Francisco, California, in 1975. The growth during the last decade can be visualized from the fact that during 1971, the total world-wide production of electricity from geothermal resources was estimated to be 700 MW, while today The Geysers installation in California, U.S.A., alone produces 750 MW of electric power. Most of the recent developments have been reported in research papers published in research journals and proceedings of various symposia. In the absence of a modern text covering these developments, it is difficult for one to learn about the assessment, exploration and exploitation of geothermal resources. The present text will fill this gap.

This book summarizes the present-day knowledge on geothermal resources, their exploration and exploitation. The first chapter draws attention to the ever-growing world-wide consumption of energy and comments briefly upon new methods of power generation. Special attention is given to the existing and possible future potential of geothermal energy including an assessment of geothermal resources in the United States. The second chapter deals with the earth's structure, temperature and heat flow within the earth, and plate tectonics hypothesis. In the third chapter the basic concepts of heat transfer are discussed and quantitative relations for the heat flow from some geometrically simple bodies are given. The fourth chapter explains the classification of different kinds of geothermal systems. In the fifth chapter different exploration techniques, viz. geological and hydrological, geochemical, geophysical and airborne, as practiced in geothermal exploration, are presented. The sixth chapter deals with the assessment and exploitation of the geothermal resources. The topics addressed in this chapter include drilling technology, reservoir physics and engineering, and production technology. The Cerro Prieto geothermal field, Mexico, is one of the very successful geothermal projects generating electrical power for which relatively complete information is available. It is discussed in detail in the seventh chapter. Chapter 8 presents the world-wide status of the geothermal resource utilization as it exists at the time of the writing of this book. It is likely that some geographical locations are not adequately covered due to lack of information. In the last, the ninth, chapter, other important topics are addressed, such as the artificial simulation of geothermal systems, the first geopressured geothermal well, environmental factors and waste disposal, reinjection of geothermal fluids, and the legal and the institutional aspects of geothermal development. The coherent presentation of these topics in a single volume shall serve as a text for graduate students as well as an introductory reference to the field for explorationists and investors. An abundant bibliography, given at the end, shall be helpful to readers in referencing more detailed documentation when so desired.

Numerous researchers, research organizations, and companies involved in geothermal resource development have been directly or indirectly helpful in preparing the text presented here. They are too many to list. Discussions and encouragement given by Professor Jim Combs during his tenure as Director of the Center for Energy Studies, University of Texas at Dallas, have been extremely useful. Much of the material was collected during extensive visits to geothermal fields and installations in the U.S.A., Japan, Italy and Mexico. Help and hospitality of concerned individuals and organizations is acknowledged. Interaction with Professor J. Singh and Dr. R.N. Singh was useful. Professors Ron Ward and Carlos Aiken offered some useful suggestions during the final stages of production of this book. Assistance of Karen Bellecci, Mike Hill, D.D. Singh, K. Suryaprakasam, Navin Chandra, K. Ramana Rao, Chandra Pal, and Del Hunt in the preparation of the material for the production of the book is gratefully acknowledged.

In the end, I would like to thank my wife Manju and my daughters Nidhi and Benu for their continuing support of my activities throughout the years.

Dallas
May, 1980

HARSH K. GUPTA

CONTENTS

Chapter 1

THE ENERGY OUTLOOK

INTRODUCTION

With the increase in world population, industrialization and improvement in the standard of living, there has been a continuously increasing consumption of energy. In the absence of substantial historical data to estimate world population prior to the seventeenth century, circumstantial evidence is often used. It is generally believed that prior to 8000 B.C., agriculture was not known and our ancestors at that time made their living by hunting and gathering. Speculating on the basis of the population densities of the hunting and gathering tribes of today, and that only one third of the world's total land area could successfully be used for such living, the world's population of 8000 B.C. is estimated to have been 5 million people. An examination of archaeological remains and recent census figures for agricultural societies suggest a world population of about 250 million people at the time of Christ, which doubled to a population of about 500 million by A.D. 1650. As a consequence of cultural, agricultural, industrial and medical revolutions, time for the population to double has reduced from about 1500 years (approximated for the period 800 B.C. to A.D. 1650) to 45 years during the period 1930—1975. Table 1-1 shows the alarmingly decreasing trend in the doubling time.

With the current rate of population growth, the United Nations forecasts that the fifth billion would be added in about a decade and the world population would exceed 6 billion much before the turn of this century.

TABLE 1-1

World population and the doubling time (United Nations, Statistical Office, Demographic Year Book)

Year	Estimated world population (in millions)	Approximate time for population to double (in years)
8000 B.C.	5	1500
A.D. 1650	500	200
A.D. 1850	1000 (1 billion)	80
A.D. 1930	2000 (2 billions)	45
A.D. 1975	4000 (4 billions)	35

In the early times, say 400,000 years ago in the "early man" era, the energy consumption of an individual was limited to what he ate (approximately 2000 calories) and this energy was obtained from the sun through vegetable and animal life. Gradually he tamed animals and used them to bear his burden of labor. Invention of the wheel improved the efficiency in the use of energy. Then man learned to harness the energy of wind and water, which became two prime movers. At about the time of the beginning of the Christian era, water and wind mills were invented and were put to a variety of uses. In the seventeenth century, coal began to be used for heating and the mining industry developed in Europe. The invention and improvement of the steam engine in the eighteenth century made power available on the spot and ushered in the industrial revolution. By the turn of the 19th century, all coal mines in England were equipped with steam engines to haul coal and men, and to pump the water. Use of these steam engines for motive power on mine railroads in 1828 was so successful that railroads spread all over the world in the next 4 to 5 decades.

Michael Faraday's discovery of electromagnetic induction in 1831 laid the foundation of the immense electrical industry of today. The invention of electric light by Edison, although it has a more profound effect on living, does not constitute a major factor in energy consumption. Then came the invention of the internal combustion engine which has proved itself to be extremely robust and reliable, and with it there has been an ever-increasing demand for oil and discoveries of oil fields.

In recent years, the increase in energy consumption has been at least as dramatic as the population explosion. Table 1-2 shows an estimate of the world's energy consumption and production for the years 1951, 1961, 1966, 1971 and 1973. These statistics have been obtained by the United Nations (World Energy Supplies Series) from the data available for approximately 185 countries with a combined population of 3802 million people in 1973. The global trend of a continuously increasing demand for energy is evident from this table. From 1961 to 1971, in a decade, world energy consumption increased by more than 80%. There are two factors contributing to this increase: (1) an approximately more than 20% increase in the world population, and (2) a world-wide improvement in the standard of living, pushing the per capita energy consumption up by almost 40%. Fig. 1-1 shows the past and projected future growth of the world population, based on the United Nations' "constant fertility projection" (with a 'medium estimate', expecting a decreasing fertility, the world population is estimated to be 6.5 billion instead of 7.4 billion by the year 2000) and of total and per capita energy consumption. Obviously, a very sincere effort needs to be made to cut down on population increase, limit our energy consumption and to find new sources of energy to maintain present standards of living in the developed countries and improve those of the developing countries.

It is interesting to note how closely per capita income and per capita energy consumption are related. The United States of America, the most

TABLE 1-2

World energy consumption and production (United Nations, Statistical Office, World Energy Supplies Series)

Year	Total production (10^6 metric tons of coal equivalent)	Total consumption (10^6 metric tons of coal equivalent)	Consumption per capita (kilograms of coal equivalent)
1951	2822	2710	1075
1961	4418	4329	1387
1966	5621	5506	1653
1971	7260	7096	1931
1973	8485	7885	2074

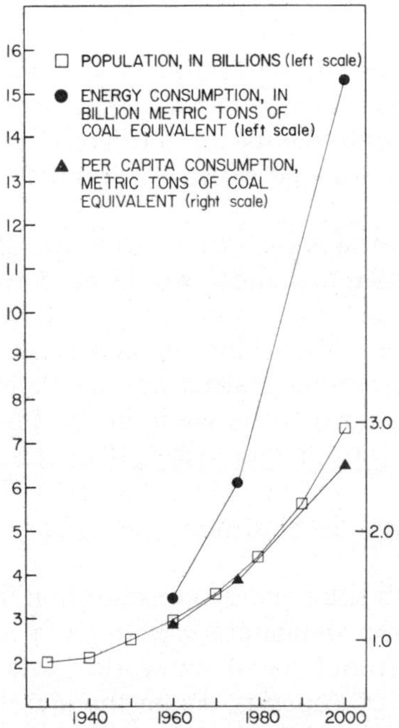

Fig. 1-1. Past and projected future growth of world population and energy consumption. The future growth of population is based on the United Nations' constant fertility projection.

affluent country in the world today, with 6% of the world's total population, consumes energy amounting to about one third of the world's total energy consumption. Fig. 1-2 shows the wide diversity in energy consumption and per capita incomes that exists today in different parts of the world. This figure also helps in visualizing the ever-growing energy demands of the future that would result from the improving standard of living in developing coun-

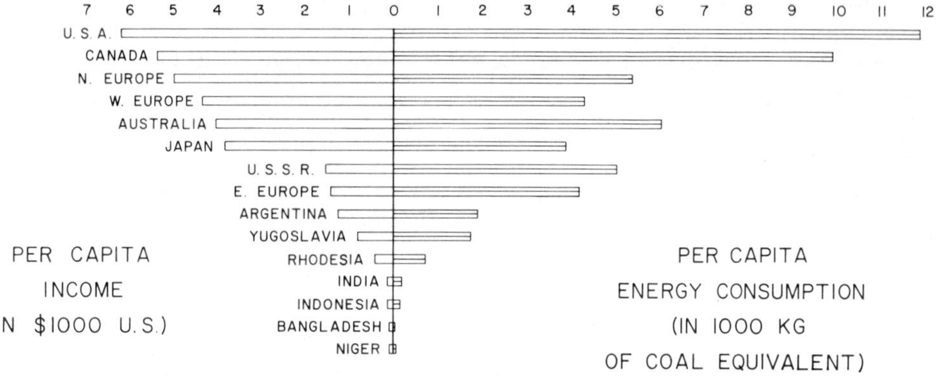

Fig. 1-2. World-wide diversity in energy consumption and per capita income.

tries, even if the developed countries control their consumption pattern and make serious efforts to hold their energy requirements at the present-day level. However, the United States' energy requirements are projected to increase by about 4%, and that of the rest of the free world by about 6% every year.

The existing energy supply and demand situation is such that very soon all the energy that could be obtained from all available sources would need to be harnessed. Especially conventional sources, such as oil and coal, would be depleted at an alarming rate. Growing concern about the environmental impact of the various energy resources places certain restrictions on their usage. To improve the energy situation, efforts need to be made in the following three directions, besides cutting the present-day increasing demands:

New methods for recovery, power generation and distribution

It is estimated that about 65% of the world's total energy consumption is provided by natural oil and gas and the world demand for petroleum is growing at a rate of 7—9% (Chilingar, 1975). About 5% of the world's total proven oil reserves exist in the United States of America. From the about 500 billion (10^9) barrels of crude oil found in the United States, 94 billion barrels have been already produced and 30 billion barrels more are estimated to be recoverable, while the remaining 376 billion barrels would be left underground. This is one typical example where technological barriers do not permit efficient utilization of a particular resource. Similar examples of inefficient extraction and power generation could be cited from practically all areas of energy-related mineral recovery from the earth, power generation and its distribution. It is hoped that scientists and engineers all over the world will exert an all-out effort to improve this situation.

Increased utilization of the limitless energy sources

A continuous flow of energy is received by the earth from the sun as direct heat and light as well as other forms of energy, such as the wind, currents and the temperature gradients in the oceans. Although the utilization of this form of energy is quite ancient, today only a minute fraction of these free energy flows is being utilized. On the contrary, mankind has exploited the finite energy resources of the earth, which have been accumulated over periods much longer than that which it is taking for man to deplete them.

However, in recent years solar energy has drawn the attention of a number of prestigious scientists and engineers who are making efforts to convert solar radiation into forms of energy more useful to mankind such as electricity, heat and chemical fuels. It is believed that within a couple of years solar-powered domestic heating and cooling plants would be commercially available at competitive prices. Enthusiastic proponents of solar-energy usage claim that in the future solar energy would become the most commonly used resource. Sørensen (1975) has outlined a plan according to which the entire energy requirements of Denmark could be provided through solar and wind energy by the year 2050. Whether Sørensen's claim is realistic or not is debatable, but there is general agreement that solar energy will play a very important role in the long run. Geothermal energy is another abundant resource of energy, and this will be discussed in the later sections.

Development and utilization of other energy sources and techniques

Today there are alternatives to replace oil. Coal, the most abundant fuel, is getting harder to mine. The way it is practiced today, coal mining is hazardous and many advanced countries are plagued by labor shortages. Automatic machinery to make deep mining safer is not available. In the absence of proper reclamation, strip mining is also not very acceptable. Burning of coal is environmentally hazardous because of sulphur contamination. The technology to remove pollutants from coal and convert it into cleaner fuels is lacking. Therefore, although enough proven coal deposits exist to meet the world's energy demands for hundreds of years, technological barriers do not permit their fullest utilization. The magneto-hydrodynamic generator, which converts heat obtained from the combustion of gases directly into electricity, is a promising development which is capable of eliminating sulphur dioxide and reducing emission of the oxides of nitrogen from coal-fired power plants.

Natural gas is one of the least expensive, least polluting and most versatile fossil fuels. Consequently over the past years, its use has been overpromoted and the reserves are depleting very fast. Its low price does not encourage oil companies to drill deeper for it rather than to drill for oil. The shortage of natural gas has revitalized efforts to produce synthetic natural gas from naphthol as well as coal. Combined gas and steam cycle systems, which combine gas and steam turbines, are important in producing clean electric power from fossil fuels.

Over the last few decades, nuclear energy has emerged as a significant addition to the traditional energy sources. It is about the only new energy technology developed commercially during the last thirty years. Extraordinary compactness is one of the unique properties of the nuclear fuels. Operating reactors use uranium fuels generating about 20,000 times as much heat as would an equivalent amount of coal. This ratio would increase to 15,000,000 times in breeder reactors now being developed. According to certain estimates, nuclear reactors would become a dominant source of heat for generating electricity by the end of this century. The U.S. Atomic Energy Commission is supporting research and development amounting to several billion dollars for the next generation of breeder reactors. Nonetheless, there is growing concern about the consequences of large-scale use of nuclear fission as a source of power as well as about the chances of a serious accident and subsequent radiation-caused damage. Long-term storage as well as disposal of radioactive wastes also cause serious problems. Natural disasters, such as earthquakes, technological failures, and deliberate sabotage, as well as human carelessness, make nuclear reactors quite vulnerable. The possibility of large-scale release of radioactive substances make nuclear-fission-generated power the most hazardous of all sources of energy.

Experimental work has demonstrated that solid organic waste could successfully be converted into synthetic fuel. This is helpful to society in two ways: (1) it provides a reusable energy source, and at the same time, (2) it solves the problem of waste disposal. For an affluent and highly wasteful society like the U.S., it is estimated that waste-converted synthetic fuel could provide 3% of the crude oil requirement or 6% of the natural gas consumption of the entire United States. However, the conversion technologies are restricted due to a scarcity of solid waste.

GEOTHERMAL ENERGY — AN ALTERNATIVE

One other potential source of energy is the earth's heat. Although hot springs have been in use for centuries for balneological purposes, the use of the earth's heat as an energy source only began early in this century when electricity was generated for the first time from geothermal steam at Larderello, Italy in 1904. By 1913, a 12.5-MW electric plant was in continuous operation there. The spread of the technology to other parts of the world had been rather slow during the first half of this century, being mostly confined to Italy. Later, interest developed in other parts of the world with intensive pioneering exploration being carried out in New Zealand, the United States of America and Japan, where electric power plants were commissioned in 1958, 1960, and 1961, respectively. Although geothermal water began to be used for space heating in Iceland in 1943, electricity production from steam started only in 1969. Much progress in utilizing this very potential source of energy has been made during the recent years, and this will be discussed in the later sections.

To appreciate the potential of geothermal energy, it may be interesting to learn that, as pointed out by Armstead (1973), through the cooling of our planet by 1°C, enough heat would be released to keep the whole world supplied with all its present-day electrical power needs for some forty million years. Similarly, it has been approximated, assuming an average heat flow of 1.2 μcal/cm^2 s over the earth's surface, that during the last one billion years (the earth's age is estimated to be 4.5×10^9 years) a total heat in the amount of 4×10^{10} cal/cm^2 has flowed out, which would have been realized through 4 tons of coal being burnt under each square centimeter of the earth's surface — that is, the complete burning of a 20-km-thick layer of coal (Bullard, 1973). This vast source of energy is now beginning to be considered as a potential future supplier of global energy requirements. Besides generating electricity, geothermal fluids are also being successfully used for space heating and cooling and other industrial and agricultural usages.

Although geothermal energy was used to generate electricity some seven decades ago, the free availability of other fossil fuels like oil and coal was not conducive to its rapid development, and substantial technical problems still remain to be solved. Today, besides being used in over a dozen countries to generate electricity totaling a few thousand megawatts, geothermal energy is being used in several other countries for space heating, agricultural and industrial purposes, and many more countries are actively exploring and assessing their geothermal resources to meet their energy requirements. Obviously, the future use of geothermal energy would very much depend on overcoming the technical barriers, both in production and utilization, and on its economic utility as compared to the other energy sources. Nonetheless, it is instructive to study the current assessment which has been made in the United States of America.

GEOTHERMAL RESOURCE ASSESSMENT FOR THE UNITED STATES OF AMERICA

During the past decade several estimates, differing by several orders of magnitude, have been made of the United States geothermal resources; e.g., White (1965), Grossling (1972), Rex and Howell (1973), Hickel (1973), National Petroleum Council (1973); the latest being by the U.S. Geological Survey (1975). As has been pointed out many times (for example, by White and Williams, 1975), these estimates vary considerably. The factors contributing to these variations are differing assumptions on the economic and technological conditions, besides the absence of adequate understanding of the extent and the nature of the resources themselves. In the U.S. Geological Survey (1975) assessment, the following terminology is used: *geothermal resource base* includes all of the stored heat above 15°C down to 10 km depth under the entire U.S.; *geothermal resources* are defined as the stored heat, both identified and undiscovered, that could be utilized using current and near-current technology, and they are divided into the following three categories: (1) *submarginal geothermal resources*: from which energy could be recovered at a cost exceeding twice the cost of other energy systems, (2) *paramarginal geothermal resources*: recoverable at costs varying from one to two times those of competitive resources, and (3) *geothermal reserves*: these are identified resources recoverable at costs competitive with other existing energy systems. The three major categories of the resource base are shown in Table 1-3 (adapted from Renner et al., 1975). Under favorable hydrological conditions, circulating water and steam transfer the geothermal energy from depth to near surface, and this system is known as the *hydrothermal convection system*. The *hot igneous systems* are caused by the upward intrusion into the shallow crust of the molten magma generated in the deeper parts of the earth's crust and mantle. Under suitable environments the heat is transferred from the deep, hot interior by the process of conduction, besides which a small fraction is contributed by the radioactive substances present in the rocks, causing the *regional conductive environments*. There is a huge quantity of heat stored under this third category (Table 1-3) in spite of its low temperature, since the area and the volume involved are very large. These different systems are discussed in detail in the next chapter.

TABLE 1-3

Estimated heat content of the geothermal resource base for the United States of America (modified from Renner et al., 1975)

Type of system	Identified systems (in equivalents of 10^9 barrels of oil)	Identified and estimated for undiscovered systems (in equivalents of 10^9 barrels of oil)
Hydrothermal convection systems	500	2000
Hot igneous systems	17,000	69,000
Regional conductive environments	5,520,000	5,520,000
Overall total	5,537,000	5,591,000

Recoverability from U.S. geothermal resources

The hydrothermal convection systems of the western United States and the geothermal sedimentary environments of the Gulf Coast account for most of the geothermal energy recoverable with present or near-future technology and prices. Estimates made by Natheson and Muffler (1975) for hydrothermal convection systems are presented in Table 1-4. The identified conductive high-temperature (>200°C) systems have an estimated electrical production potential of 8000 MW-cent* (Natheson and Muffler, 1975)

*1 MW-cent is equivalent to 1000 kW produced continuously for 100 years.

TABLE 1-4

U.S. hydrothermal convection systems considered recoverable with the present and near-future technology and disregarding cost (modified from Natheson and Muffler, 1975)

Type of system	Heat in ground[1]	Heat at the wellhead[1,2]	Conversion efficiency	Beneficial heat[1,3]	Electrical energy (MW-cent)[4]	Total electricity for 30 years[5] (MW)
High-temperature systems (>150°C; for electricity generation)						
Identified resources	180	40	0.08—0.2			
Reserves					3500	11,700
Paramarginal resources					3500	11,700
Submarginal resources					> 1000[6]	> 3300[6]
Undiscovered resources	830	200	0.08—0.2		38,000[7]	126,700[7]
Intermediate-temperature systems (90—150°C; mainly non-electrical uses)						
Identified resources	240	60	0.24	15		
Undiscovered resources	710	180	0.24	40		
Total	1960	480		55	46,000	153,400

[1] Heat content in equivalent of 10^9 barrels of oil.
[2] Assumed recovery factor of 0.25 for all convective resources.
[3] Assumed that the energy would be used on location.
[4] 1 MW-cent is equivalent to 1000 kW produced continuously for 100 years.
[5] Assumes that each MW-cent of electricity can be produced at rate of 3.33 MW for 30 years.
[6] Small because exclusion of systems with temperatures below 150°C.
[7] Probably 60% will be reserves and paramarginal resources. Costs of discovery and development are more speculative than for identified resources.

TABLE 1-5

U.S. geothermal resources of geopressured sedimentary environments considered recoverable with present and near-current technology and disregarding cost (modified from Papadopulos et al., 1975)

	Heat in pore fluids[1]	Heat equivalent at wellhead[1,2]	Electrical energy (MW-cent)[3]	Total electricity for 30 years[4] (MW)
Gulf Coast geopressured fluids in sediments of Tertiary age to depths of 7 km assessed for onshore parts only	7500			
Plan 1, maximizes total recovery over 20-year period; no pressure decline below 2000 psi; 17,160 wells; subsidence estimated 5—7 m				
Total of thermal, methane (thermal equivalent) and mechanical (thermal equivalent) energy		250	34,350[5]	114,490[5]
Plan 2, assumes drawdown to land surface, unrestricted subsidence, and mechanical energy not utilized; 25,850 wells; estimated average subsidence 7 m				
Total thermal and methane (thermal equivalent) energy		380	38,140	127,100
Plan 3, limits estimates with average subsidence to 1 m; 4000 wells				
Total of thermal, methane (thermal equivalent) and mechanical (thermal equivalent) energy		60	9250	30,900
Other unassessed parts of Gulf Coast geopressured environments, onshore and offshore to 10 km depth[6]	15,000	350	50,000	166,700
Other geopressured environments to 10 km depth[6,7]	7500	180	25,000	83,300

[1] Thermal energy equivalent in 10^9 barrels of oil.
[2] All plans assume 0.15 m^3/s flow rate per well and saturation of water with methane, but reliable data are lacking.
[3] 1 MW-cent is equivalent to 1000 kW electrical energy produced continuously for 100 years.
[4] Estimates made for production period of 20 years converted to 30 years for consistency with Table 1-4.
[5] In parts reserves but mostly paramarginal, depending on environmental and other costs.
[6] Thermal equivalent of methane included in heat at wellhead but excluded from electrical energy; recoverable part highly speculative because of unknown porosities and permeabilities, but probably largely submarginal.

under certain reasonable assumptions. The intermediate-temperature (90–150°C) conductive resources are presently submarginal for the generation of electricity, but under certain conditions are suitable for space heating and industrial uses. It is expected that the availability of a relatively inexpensive source of energy would attract industry, since the over-all efficiency of the direct use of the geothermal energy for heating is more than that for generating electricity for the same purposes. The total heat energy available for such local industrial and space heating use is estimated to be equivalent to about 15 billion barrels of oil from the identified systems.

Estimates have been made by Papadopulos et al. (1975) for the heat content of pore fluids belonging to the onshore geopressured Gulf Coast to depths up to 7 km. Their estimates are shown in Table 1-5. Under different plans, numbered 1 through 3, and assuming that the mechanical energy is convertible into electricity with 80% efficiency and that the thermal energy at reservoir temperatures is convertible into electricity with 8% efficiency, the available energy could vary from 9250 to 38,100 MW-cent of electricity. This assessment of recoverability of electrical energy from the geopressured geothermal resources of the Gulf Coast does not consider the cost factor.

In the absence of necessary recovery technology, the waste heat content of magma, hot dry rock near volcanoes and other conduction-dominated parts of the resource base are not considered as geothermal resources. Lately, much attention is being paid to developing suitable recovery technology to utilize the unlimited amount of energy contained in the hot dry rocks.

The statistics given and statements made are based on the data available up to June, 1975 as used by the U.S. Geological Survey (1975) in their assessment of the geothermal resources of the United States. In their concluding remarks, White and Williams (1975) mention that disregarding cost, the total magnitude of the geothermal resources which could be exploited by present or near-future technology is very large. The total production potential of the identified high-temperature convection systems and the onshore parts of the geopressured-geothermal field is estimated to be 42,000 MW-cent of recoverable electricity. This amount of electricity is equivalent to that generated by 140 Hoover Dams or 140 average modern nuclear power plants. The unassessed geopressured part of the Gulf Coast and other sedimentary basins, together with the undiscovered convection systems, is estimated to have a potential of at least 100,000 MW-cent and probably more than half of this energy could be harnessed at prices with upper limits no more than twice the present costs.

Chapter 2

BASIC CONCEPTS

INTRODUCTION

In the previous chapter the world's growing demands for energy have been discussed and it has been demonstrated that in the future, with proper development, geothermal resources could make an important contribution to this requirement. We shall now briefly discuss the structure of the earth, its thermal history and heat flow and the processes going on within it which are responsible for the generation and transportation of heat.

EARTH'S STRUCTURE

The earth's structure can be approximated by a series of concentric spherical shells. The large-scale features of the earth's internal structure are shown in Fig. 2-1. The core, constituted by the two innermost regions, has the greatest average density, exceeding 10 g/cm^3. In spite of differing views on the details of the composition of the core, it is now fairly well accepted that iron-nickel alloy is the most likely constituent. However, the observed characteristics of the core do not match with its being purely iron-nickel — it is too dense, and the presence of some lighter material is postulated. Silicon has been proposed to be an alloying element in the core. For this to be true, it is necessary to assume the presence of suitable conditions, early in the time of the earth's formation, making it possible for large quantities of silicon to reduce — stripping away oxygen atoms and adding electrons. Sulphur is another light element which has been recently suggested as being present in the core. This would require quite different conditions to exist in the early history of the earth. The scope of this book does not permit discussion on the origin of the core. Whatever may be its origin, certain aspects of the present structure of the core are well established from seismological evidence. The outer part is molten since it does not transmit shear waves. The study of compressional waves, that travel through the inner part of the core, show higher velocities leading to the suggestion of the inner core being solid. Results of the study of free oscillation of the earth, as well as the detection of seismic waves that have travelled through the inner core as shear waves confirm the above suggestion.

The core is covered by the mantle with an average density of 4.5 g/cm^3 indicating that its constituents are rocky rather than being metallic. The composition of the mantle is not completely determined. However, based on its density, seismic wave velocities and study of rocks believed to have come

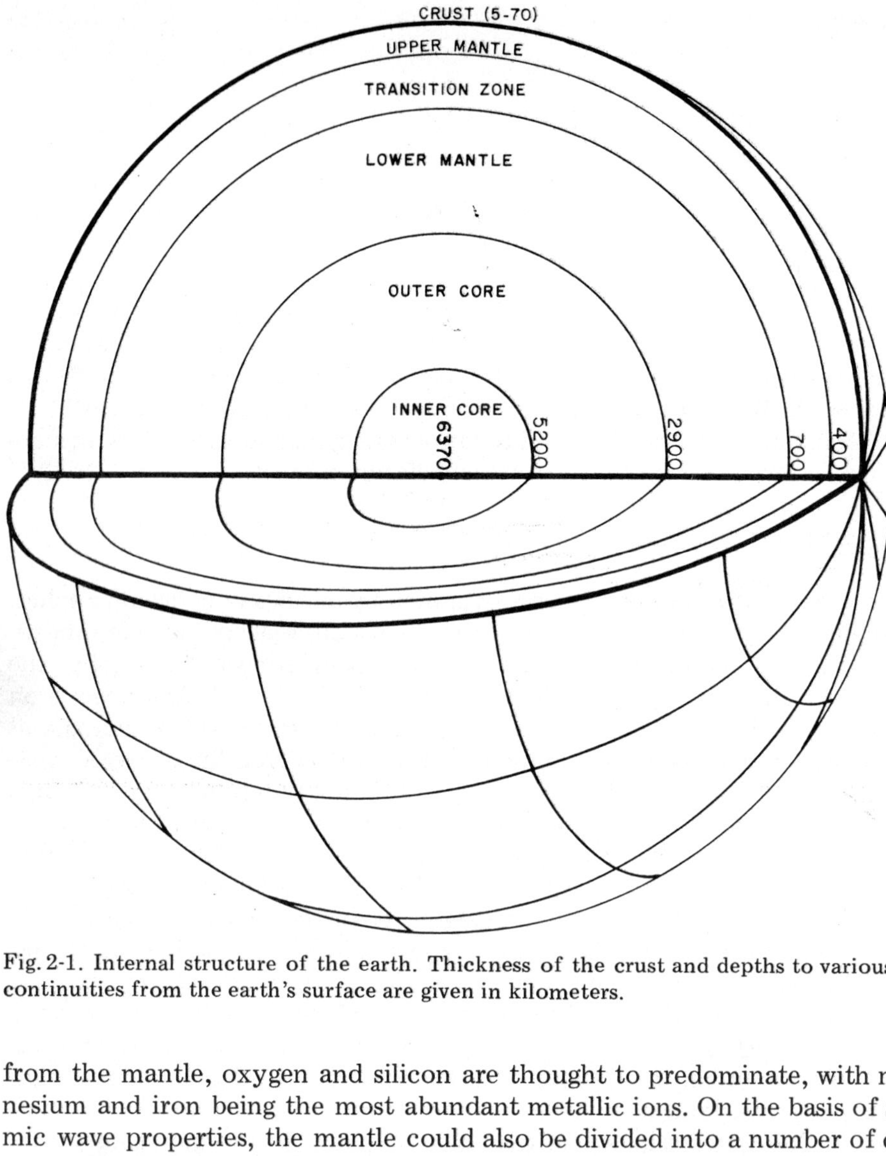

Fig. 2-1. Internal structure of the earth. Thickness of the crust and depths to various discontinuities from the earth's surface are given in kilometers.

from the mantle, oxygen and silicon are thought to predominate, with magnesium and iron being the most abundant metallic ions. On the basis of seismic wave properties, the mantle could also be divided into a number of concentric shells. The lower mantle extends from a depth of about 700 km within the earth to the top of the core at 2900 km. As a result of the increase in pressure, the seismic velocity and density increase with depth in the lower mantle. The amount of iron in the silicate minerals also increases with depth, a factor which would also contribute to increase in density and velocity.

The sudden changes in seismic velocity in the transition zone extending from about 400 km depth to the top of the lower mantle are more likely related to alterations in the crystal structure than to changes in the composition. For example, when the pressure reaches a certain value (at approximately 400 km depth), the atoms in the crystal structure of mineral olivine are rearranged into more compact form resembling the mineral spinel. With each phase change in the transition zone, density and seismic velocity increase. It is interesting to note that the deepest known earthquakes at 700 km coincide with the seismic velocity increase, probably associated with a phase change marking the boundary between the transition zone and the lower mantle.

The upper mantle extends from the base of the crust to a depth of about 400 km and is mainly composed of olivine, pyroxene and garnet. These minerals would have the same stable crystalline structure at pressures reached within the upper mantle as on the surface of the earth. The existence of an approximately 100-km-thick, low-velocity zone for both the shear and the compressional waves with its top at depths varying from 80 to 120 km is one of the remarkable features of the upper mantle. This low-velocity zone is particularly well defined for the shear waves and is more prominent in the upper mantle below the oceanic basins. The low-velocity zone is probably caused by small pockets of molten material scattered throughout the mass of solid minerals. With the increase in pressure, the temperature at which part of the mantle minerals begin to melt increases as shown in Fig. 2-2. The temperature in the mantle increases steeply with depth and can rise into the melting zone and fall again below the melting temperature at greater depths. Besides causing lower seismic velocities, the semimolten material reduces the mechanical strength considerably in this zone.

The Mohorovićić discontinuity, named after its discoverer in the early twentieth century, and also known as the Moho for short, by definition, separates the crust from the upper mantle. Seismic P-wave velocities suddenly increase from 6.5 km/s above the discontinuity to 8 km/s below it. The

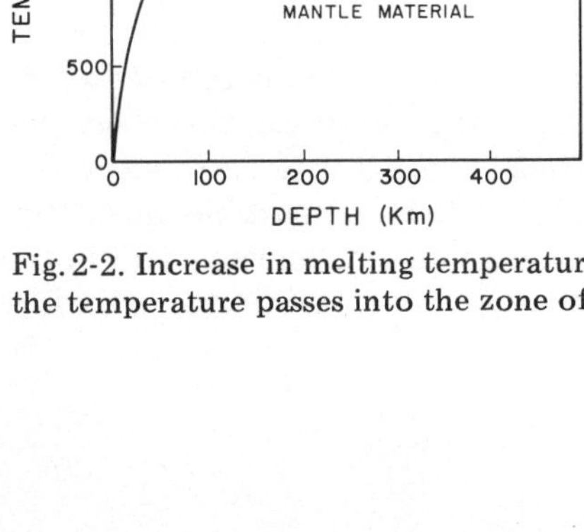

Fig. 2-2. Increase in melting temperature with depth in the earth. According to this model, the temperature passes into the zone of partial melting over a restricted depth.

Moho at places may be due to phase change or a compositional boundary or both. The composition of the crust beneath the oceans differs considerably from that beneath the continents. The oceanic crust accounts for about 65% of the total earth's surface and it is covered with an average 4 km depth of water. Below the water layer there exists about 0.5 km of sediments overlying an approximately 1.5-km-thick layer of basaltic volcanic rocks. The next 6 km, down to the Moho, mostly consist of metamorphic equivalents of the basaltic volcanic rocks and other iron- and magnesium-rich igneous rocks.

In contrast to the relatively uniform oceanic crust, the continental crust varies in thickness from less than 25 km under certain shield areas to more than 60 km below certain high mountains, with an average thickness of 35 km, and it has a complicated structure. At the surface a great variety of rocks are exposed. These include sediments, such as clays, sandstones or limestones and ancient shields, mostly composed of granites and volcanic lavas. The underlying basement is composed of granites formed either through freezing of the molten material or metamorphism of the sediments. Being a result of complicated thermal, mechanical and chemical processes, the granites vary considerably in their mineralogical composition, and are generally characterized by a low melting point and a considerable portion of free silica in the form of crystalline quartz. With the increase of depth, the proportion of basalt increases, and at many places a second-order discontinuity, known as the Conrad discontinuity, which separates the granitic layer from the basaltic layer, has been inferred from the seismic wave travel-time studies. The depth of the Conrad discontinuity varies considerably from region to region, being estimated to be from one third to two thirds of the continental crustal thickness. Below this discontinuity and extending to the Moho, a basaltic layer has been inferred.

In addition to the basic differences in the continental and the oceanic crust, such as the elevation, thickness, structure and overall composition, the transition between the oceanic and the continental crust is also very interesting. The continents extend as shallow continental shelves beyond the shore line. The total area covered by the shelves amounts to about 6% of the total global surface. The edge of the continental shelves is followed by the continental slope leading into deeper water. Deposition of sediments at the base of continental slopes gives rise to the continental rise on the edge of the deep ocean basins.

TEMPERATURES WITHIN THE EARTH AND HEAT FLOW

It has been known for a long time that temperatures in mines exceed those at the surface. This increase in temperature with depth causes the thermal gradient, resulting in the flow of heat through the earth's surface. The determination of heat flow is one of the few measurable quantities which provides information on thermal state and heat generation within the earth. At the present time direct penetration to the earth's interior is limited to 7- to 8-km-deep wells where temperatures in excess of 200° C have been measured. Although the distribution of temperature within the earth's interior is a clue to many problems, it is one of the least well understood physical properties of the earth. The uncertainty is mostly because the temperature distribution is inseparable from the hypothesis of the earth's origin and its thermal history of the last 4.5 billion years which are still much debated (Lubimova, 1967, 1969). Here we shall not discuss the various hypotheses and shall only present temperatures within the earth according to a certain model (Fig. 2-3). In the deeper portions of the earth the actual temperatures could deviate from those shown in Fig. 2-3 by as much as 1000°C.

The number of places where both the temperature gradient and the thermal conductivity have been measured in boreholes, has increased from a few hundreds in the early sixties to a few thousands at the present time. The heat flow values (product of thermal conductivity and the thermal gradient) at the surface of the earth average about 1.2 μcal/cm^2 s and most of the variation in heat flow lies within a factor of three about the mean (Chapman and Pollack, 1975). A certain pattern in the heat flow values, related to the tectonic elements has been observed. In general, the continental heat flow values are found to be correlated with the age of the last tectono-thermal mobilization and the ocean heat flow with the age of the ocean floor (Fig. 2-4).

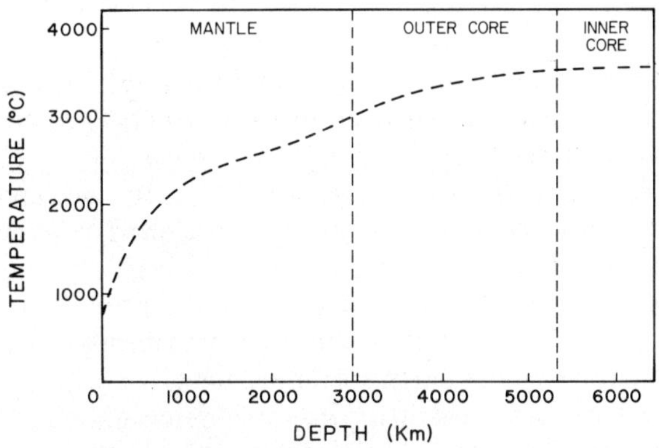

Fig. 2-3. Variation of temperature with depth within the earth. In deeper portions, temperatures could deviate by as much as 1000°C.

The source for heat flow through the earth's surface

The original heat contained in the earth at the time of its formation and the radioactivity seem to be responsible for the temperatures within the earth and the observed average heat flow of 1.2 μcal/cm^2 s at the earth's surface. Since the initial temperatures at the time of the formation of the

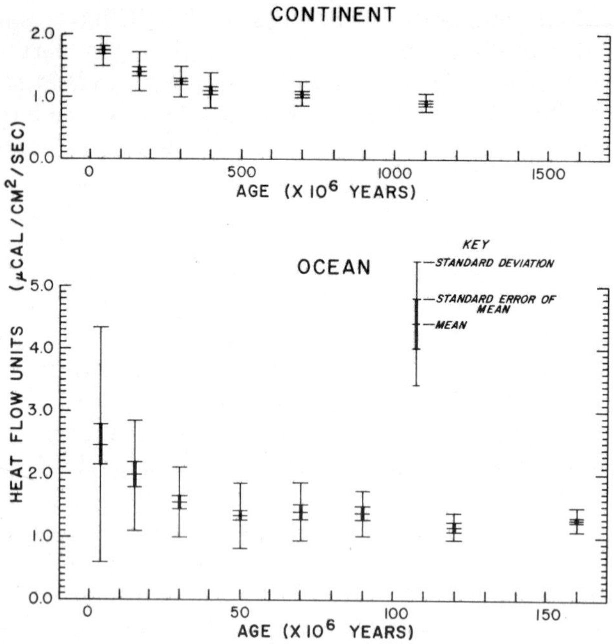

Fig. 2-4. Observed variations in continental and oceanic heat flow values and their dependence on the age of the rocks (after Chapman and Pollack, 1975).

earth are not reliably estimated, it is difficult to determine the contribution of the earth's original heat to the near-surface thermal state of the present day. However, as shown for the first time by Lord Kelvin, the earth is much too young for the initial thermal conditions in its central region to have any manifestations at the surface. Although it needs to be mentioned that in Lord Kelvin's model heat was considered to be transferred through conduction only and his argument would need modification if mantle convection were to be considered, still it seems unlikely that the present-day heat flow at the surface of the earth is affected by the earth's original heat.

The amount of radioactive substances present in the rocks releases enough heat to account for a major portion of the total heat flow observed at the earth's surface. Of all the radioactive elements present, significant heat contributions are made by uranium (^{238}U and ^{235}U), thorium (^{232}Th) and potassium (^{40}K) and their products. The two isotopes of uranium, as well as thorium, decay to lead through many intermediate stages. ^{40}K, the rare isotope of potassium, decays through two routes to calcium and argon. Pertinent details of half-life and heat production by the most common radioactive elements in rocks are given in Table 2-1. As has been pointed out in the past (for example, Bullard, 1973), the total observed heat flow at the earth's surface can easily be accounted for by the radioactivity of the earth's crust. It is difficult to assign exact values of average concentration of radio-

TABLE 2-1

Major abundant radioactive elements found in rocks, their half-lives and heat generation

Isotope	Half-life (years)	Heat production (cal/g yr)
^{238}U	4.5×10^9	0.71
^{235}U	0.71×10^9	4.3
^{232}Th	13.9×10^9	0.20
^{40}K	1.3×10^9	0.21

active elements in the crust. For the sake of calculation, let us presume the average concentrations to be 2 ppm for uranium, 8 ppm for thorium and 2% for potassium in the continental crust. Such a composition would release 3.6×10^{-6} cal/g yr of heat and a 39-km-thick crust would account for the observed average 1.2 μcal/cm^2 s surface heat flow. It is also necessary to assume that the material below the crust is much less radioactive or otherwise the heat flow at the earth's surface in the continent would have been much more than that observed. Measurements of radioactivity on rock samples, which are considered to have come from the mantle, have revealed very weak radioactivity. Beneath the ocean floors, things have to be different. The average heat flow over the continents and oceans is practically the same, although ocean floors are underlain by only about 5 km of feebly radioactive basalt to the Moho, and hence, most of the heat must come from the mantle.

VOLCANOES AND PLATE TECTONICS

As discussed earlier, the temperatures within the crust are much below the melting point of rocks and hence, normally no molten material should exist close to the surface of the earth. Volcanoes are very special features which mostly occur around the island arcs and the mid-oceanic ridges. The recent conceptual development of the theory of plate tectonics, and its successful application in explaining the observed geophenomena, provide an acceptable explanation for the restricted geographical distribution of the volcanoes. In the following, we shall briefly review the plate tectonics theory to understand the occurrence of volcanoes and other regions of high heat flow.

The two basic assumptions of plate tectonics are very simple. The first is to assume that the outer portion of the earth consists of a few large rigid plates, very much like caps on a sphere, which do not undergo any significant internal deformation. The second is to assume that each plate is in relative motion with respect to the other plates and that significant deformation takes place only at the plate boundaries. Plates are areas within which horizontally directed tectonic processes of any significance are not at present occurring. There could be small vertical movements within the plates. However, they would be of much smaller magnitude compared to large horizontal movements at the plate boundaries. The plate boundaries are mostly marked by narrow belts of seismic activity. The geographical distribution of the major plates is shown in Fig. 2-5.

The outer portion of the earth, including the crust and extending to the low-velocity zone in the upper mantle, which constitutes the rigid plates, is known as the lithosphere. The portion lying below the lithosphere, and probably extending to 700 km depth, is called the asthenosphere. In contrast to the rigid lithosphere, material in the asthenosphere is hot enough to deform and flow internally.

The processes within the earth's interior can make the plates move apart leaving a gap between them, or move plates towards one another, shortening and/or buckling them, as well as making them slide past each other at their contact. With the receding of plates from one another, the hot and partially molten material from the asthenosphere rises to fill up the gap, resulting in sea-floor spreading. Gradually this semi-molten material cools down, becoming a part of the rigid plates on the two sides of the spreading zone. It appears that equal amounts of the new material are added on the two sides, and hence, the spreading is usually symmetrical. The newly added material takes time in cooling and shrinking, and the new lithosphere gradually moves away from the zone of spreading. In the meantime more of the lithosphere is

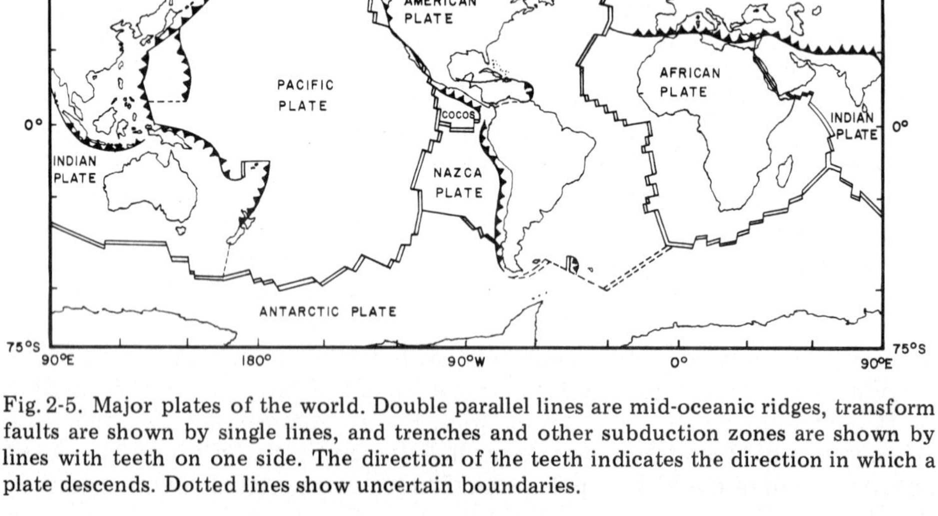

Fig. 2-5. Major plates of the world. Double parallel lines are mid-oceanic ridges, transform faults are shown by single lines, and trenches and other subduction zones are shown by lines with teeth on one side. The direction of the teeth indicates the direction in which a plate descends. Dotted lines show uncertain boundaries.

added, causing the center of oozing to be elevated above the surrounding older sea floor. The elongated elevated zones thus created in the center of the sea-floor spreading are called mid-oceanic ridges, since they occupy central positions in the Atlantic and Indian Oceans. Besides their elevation, the other physical manifestation of the mid-oceanic ridges are the associated linear belt of earthquakes characterized by a normal faulting-type focal mechanism, high heat flow and a typical symmetrical pattern of the observed magnetic anomalies. The spreading rates of the diverging plates at the mid-oceanic ridges are estimated to vary from 2 to 20 cm/yr, the average being 6 cm/yr.

The processes that create and modify the continental crust are much more complicated compared to the generation of the oceanic crust. Since the surface area of the earth seems to remain constant, it is implied that the creation of the lithosphere at divergent plate boundaries in the midoceanic ridges must be balanced by the destruction of the lithosphere at convergent plate boundaries. This shortening takes place through either the deflection of one of the plates into the asthenosphere — a process known as subduction, or through the crumpling and buckling of the two colliding plates. As mentioned earlier, the oceanic lithosphere is derived directly from underlying mantle material, and hence, its overall composition must be similar. Cooling makes it denser and therefore easier for it to sink back and be resorbed in the mantle. When two oceanic plates collide, one of them sinks and goes back to the mantle. At the collision of an oceanic plate with a continental plate, the oceanic plate is consumed in the subduction zone. In contrast with the oceanic lithosphere, the continental lithosphere is not easily forced down and consumed in the subduction zone since it includes the continental crust which has a different composition from the mantle material and is much less dense. Therefore, when two continental plates collide, neither of them is consumed, and buckling and crumpling take place, resulting in great mountain ranges, such as the Himalaya. Fig. 2-6 shows typical subduction zone models.

Oceanic trenches are distinct topographic features associated with the subduction of the oceanic lithosphere. The oceanic plate bends down over a distance of the order of 100 km as it enters the subduction zone, straightens out again and penetrates the asthenosphere at a steep angle. The contact between the overlying and the sinking plate is down at the bend causing the trench (Fig. 2-6). The deepest known trenches in the western Pacific are more than 10 km deep.

It has been estimated that at places the downgoing lithospheric plate in the subduction zone can retain its identity to up to 700 km depth when it finally warms up and mixes up with the asthenospheric material. This process involves a period of the order of 10 million years. The descending slab of oceanic lithosphere could cause additional melting in the mantle or the upper oceanic crust part of the descending lithosphere itself melts in the 150- to 300-km depth range. The molten material rises through the overlying

A. COLLISION OF TWO PLATES, BOTH WITH OCEANIC LITHOSPHERE

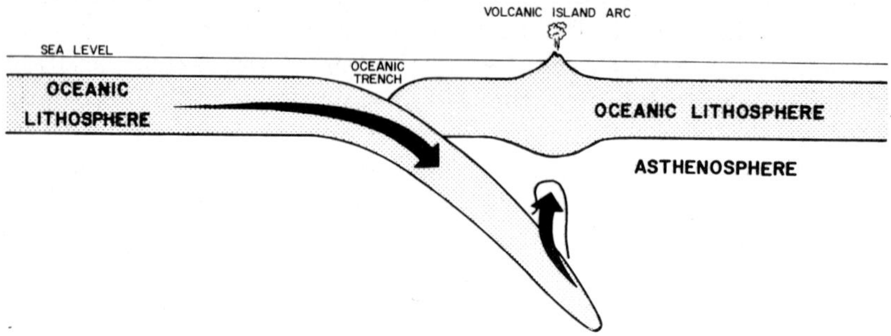

B. COLLISION OF TWO PLATES, ONE OCEANIC AND ONE CONTINENTAL

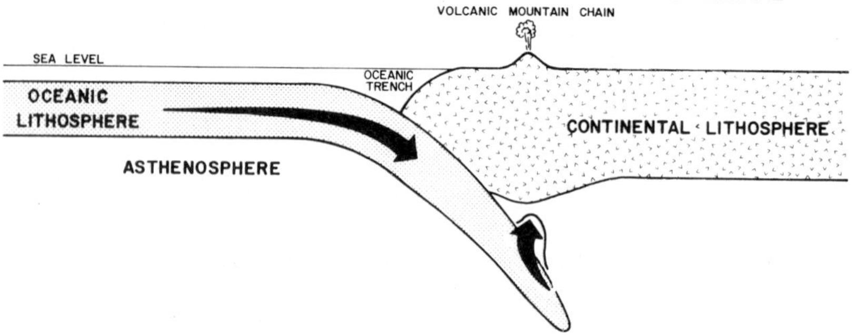

C. COLLISION OF TWO PLATES, BOTH WITH CONTINENTAL LITHOSPHERE

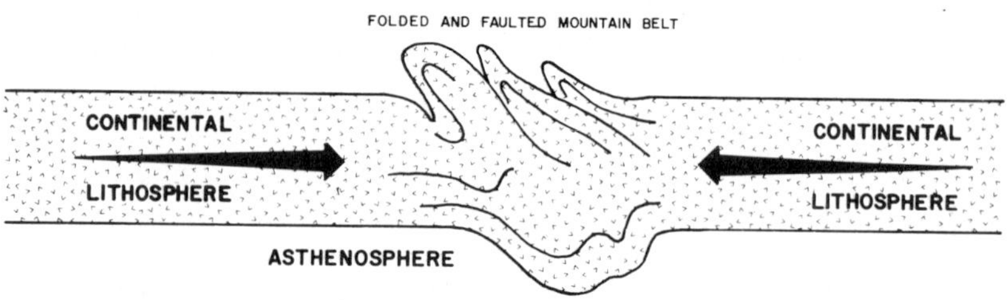

Fig. 2-6. Converging plates and the resultant physiographic features.

plate and breaks through the earth's surface to form volcanoes. This explains the occurrence of most volcanoes over the subduction zone (Fig. 2-7). When the overlying lithosphere, through which the molten material rises, is oceanic, the volcanoes build island arcs. When the overlying lithosphere is continental, volcanic mountain chains are created.

The two major earthquake belts, namely the circum-Pacific belt and the Alpide belt, which account for most of the earthquakes, lie in the zones of subduction and continent-continent collision, respectively. Earthquakes associated with the subduction zones can distinctly be divided into four

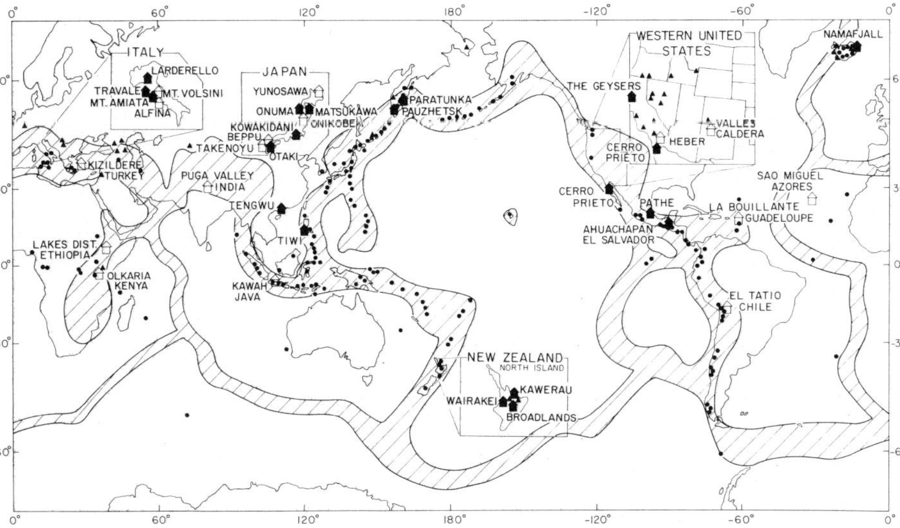

Fig. 2-7. World-wide seismic belts (hatched areas); volcanoes (filled dots); geothermal fields under production (filled huts), and being developed (open huts); and geothermal prospects (filled triangles).

zones. In the outer zone the earthquakes originate within the oceanic crust or the upper lithosphere of the sinking plate near the outer wall of the trench. The shallow zone includes earthquakes occurring at the contact of the two plates extending in depth to 100 km. These are followed by the intermediate-depth earthquakes in the range of 100—300 km, and they occur in the downgoing slab rather than at its contact with the asthenosphere. Deep-focus earthquakes occurring in the 300- to 700-km depth range are not found in all subduction zones. All these earthquakes are related to the stiffness of the sinking lithosphere and hence indicate its penetration through the asthenosphere.

Earthquakes associated with the continent-continent plate collision are much more diffused and obviously do not show any pattern analogous to those observed at the oceanic plate-continental plate collision.

As can be noted in Figs. 2-5 and 2-7, plate boundaries, earthquakes, volcanoes and regions of high heat flow where geothermal fields are located are all related.

Chapter 3

HEAT TRANSFER

INTRODUCTION

Heat transfer has been the subject of a number of text books, e.g., Jacob (1964), Isachenko et al. (1969) and others. Here we discuss briefly the fundamentals relevant to the present text. Of the three possible modes of heat transfer, namely conduction, convection and radiation, conduction governs the thermal conditions in almost entirely solid portions of the earth, convection dominates the thermal conditions in the zones where large quantities of fluids (mostly molten rocks) exist, and the temperatures at the earth's surface are controlled by the process of heat exchange between the sun and the earth through radiation.

TEMPERATURE, HEAT AND ITS STORAGE

A body is said to be warmer, or at a higher temperature, when it spontaneously gives out its heat to another body. The inherent sensing capability of the skin of human and animal bodies gave rise to the concept of temperature. With the increase of temperature, the disordered kinetic energy of the molecules constituting a body increases and hence in physics, temperature is a measure of the disordered kinetic energy of the molecules. Temperature is not expressed in basic units (length, mass and time); it is mostly expressed in degrees and three scales, based upon the boiling and freezing temperatures of pure water, are in vogue. These are the Celsius (°C), the Fahrenheit (°F) and the Kelvin (°K) scales with the following correspondence:

$$(°K) = (°C) + 273,$$
$$(°C) = 5/9\,[(°F) - 32]$$
$$(°F) = 9/5\,(°C) + 32$$

Heat is a form of energy which originates from other kinds of energy and under suitable conditions can be transferred into still other forms of energy. When subjected to heating, a body consumes the heat energy through the increase of its internal energy and work done in doing so. The calorie is the basic unit of heat energy and is equal to the heat required to raise the temperature of 1 cm^3 of pure water through 1°C at normal atmospheric pressure.

Change in the temperature of a body is accompanied by a change in the heat content of the body, also known as enthalpy. The amount of heat that a body is able to store depends upon its specific heat, which is defined as the amount of heat required to raise the temperature of one gram of the material by 1°C. The amount of heat, Q, required to raise the temperature from θ_1 to θ_2 of a body of mass, m, and specific heat, c, is given by:

$$Q = m \cdot c(\theta_2 - \theta_1) \tag{3-1}$$

The change in heat content, or enthalpy, is conventionally understood to be taking place at a constant volume. In the case of gases, it is necessary to distinguish between change in enthalpy at constant volume and those at constant pressure. Since the coefficient of thermal expansion (change in volume of 1 cm^3 of material per °C) for solids and liquids is very small, the abovementioned distinction is of no practical significance. When the specific heat is strongly temperature dependent, equation (3-1) could be written as:

$$c = \frac{1}{m} \cdot \frac{\partial Q}{\partial \theta} \tag{3-2}$$

In geothermics, one invariably has to deal with fluid-filled porous rocks. The specific heat of a rock with a_1 grams of water and a_2 grams of dry-rock content could be expressed as:

$$c_{\text{wet}} = \frac{a_1 c_{\text{water}} + a_2 c_{\text{dry}}}{a_1 + a_2} \tag{3-3}$$

or simply by:

$$c_{\text{wet}} = (1 - a_1) c_{\text{dry}} + a_1 c_{\text{water}} \tag{3-4}$$

when $(a_1 + a_2) = 1$, i.e., one gram of wet porous rock containing a_1 grams of water. The specific heat of dry rocks and soils is found to be of the order of 0.2 cal/g °C. With the increase of water content, the specific heat of the porous rocks increases.

In addition to the temperature changes, a change in phase also alters the enthalpy. A phase change involves either absorption or release of heat. The amount of heat absorbed or released during the phase change of a unit mass

TABLE 3-1

Thermal properties of water and some selected elements

Material	Melting point (°C)	Boiling point (°C)	Latent heat of melting (cal/g)	Latent heat of vaporization (cal/g)
Water	0	100	79.4	539.1
Iron (pure)	1530	2500	65	1520
Gold	1063	2700	16	420
Carbon	3540	4000		12000
Platinum	1773	3800	27	600
Sulphur, rhombic	112.8	444.6	9.4	70
Silicon	1410	2350		3360

of a substance, the temperature being constant, is called the latent heat. In geothermics, a change in enthalpy associated with the change of temperature is known as the free heat. However, one usually faces situations where both temperature and phase changes are involved. This could be expressed as:

$$\Delta Q = \Sigma cm \ \Delta\theta + \Sigma L \ \Delta m \tag{3-5}$$

where L is the latent heat and the summation is carried out for a system constituted by materials of differing specific and latent heats. Thermal properties of some commonly observed elements are included in Table 3-1.

HEAT CONDUCTION

Inside a closed solid body (no heat permitted to enter or leave the body) the temperature differences are equalized with the passage of time through the process of interaction of the molecules. All the molecules attain a certain energy level and the temperatures are same all over the body. The process of attaining the same temperature all over the body is associated with the flow of heat. The rate of heat flow between two points is directly proportional to their temperature difference. We shall now investigate this mode of heat transformation for isotropic as well as anisotropic solids.

Isotropic solids

Consider a rectangular coordinate system, and assume that at a certain time a change in temperature is taking place only along the x-axis, the direction of heat flow being in the positive direction of the x-axis. Let the temperature at a distance x from the origin of coordinates be θ. Since we have assumed that heat is flowing in the positive x direction, the temperature would decrease in that direction. Let the temperature at a distance $(x + dx)$ be $(\theta - d\theta)$. The quantity $d\theta/dx$ is called *temperature gradient* and the amount of heat, Q_x, flowing through a plane surface of area F at right angles to the x-axis in time T is proportional to the temperature gradient, i.e.:

$$Q_x \propto -\frac{d\theta}{dx} FT \tag{3-6}$$

or:

$$Q_x = -K\frac{d\theta}{dx} FT \tag{3-7}$$

where K is a physical quantity known as the *thermal conductivity*, which depends on the nature of the material through which the heat is flowing and is affected by physical conditions such as the temperature. The quantity of heat flowing per unit time per unit area ($T = F = 1$), q_x, is called *heat flow*:

$$q_x = -K \frac{d\theta}{dx} \tag{3-8}$$

Equation (3-8) is known as Fourier's equation. Let us now consider the general case of temperature change from a point to another point in an isotropic solid. The temperature at any given time t at point P could then be expressed as:

$$\theta = f(x, y, z, t) \tag{3-9}$$

At a given time, a surface in the solid which has the same temperature is called an isothermal surface. Since at a given time, no point can have two temperatures, isothermal surfaces never cut one another. Let us now find the heat flow at this point P, across any surface, and not necessarily the isothermal surface through P. Let the xy plane be taken tangentially to the isotherm at P. Then the fluxes across the planes through P and parallel to the coordinate axes are:

$$q_x = -K\frac{\partial \theta}{\partial x}, \quad q_y = -K\frac{\partial \theta}{\partial y}, \quad q_z = -K\frac{\partial \theta}{\partial z} \tag{3-10}$$

Introducing a vector form:

$$\vec{q} = -K \operatorname{grad}\theta \tag{3-11}$$

where grad θ is the temperature gradient vector representing the magnitude and direction of the steepest change in temperature, its direction being normal to the isothermal surface passing through the point under consideration. From equation (3-11) it follows that within an isotropic medium, heat flow is a vector quantity with its direction normal to the surface of constant temperature.

Anisotropic solids

In geothermics one often has to deal with media having anisotropic thermal properties. This is particularly common while dealing with the problems of heat flow in sedimentary rocks where the properties in the bedding plane (plane of sedimentation) tend to differ from properties at right angles to it. Similarly, one often has to deal with crystals and rock-forming minerals belonging to different systems, such as the monoclinic (feldspars, micas), orthorhombic (pyroxenes, olivines), hexagonal (quartzs, ilmenites) and other systems.

In an anisotropic media, the thermal conductivity could be represented by a symmetrical tensor of the second order of the type:

$$\begin{bmatrix} K_{11} & K_{12} & K_{13} \\ K_{21} & K_{22} & K_{23} \\ K_{31} & K_{32} & K_{33} \end{bmatrix} \tag{3-12}$$

Assuming each component of the heat flow vector to be linearly dependent on all components of the temperature gradient at that point, we could write the heat flow in the three mutually perpendicular directions for an anisotropic media as:

$$q_x = -\left(K_{11}\frac{\partial\theta}{\partial x} + K_{12}\frac{\partial\theta}{\partial y} + K_{13}\frac{\partial\theta}{\partial z}\right)$$

$$q_y = -\left(K_{21}\frac{\partial\theta}{\partial x} + K_{22}\frac{\partial\theta}{\partial y} + K_{23}\frac{\partial\theta}{\partial z}\right) \qquad (3\text{-}13)$$

$$q_z = -\left(K_{31}\frac{\partial\theta}{\partial x} + K_{32}\frac{\partial\theta}{\partial y} + K_{33}\frac{\partial\theta}{\partial z}\right)$$

In the case of anisotropic layers conductivity being different in three mutually perpendicular directions coinciding with the x-, y- and z-axes, equation (3-13) simplifies to:

$$q_x = -K_1\frac{\partial\theta}{\partial x}$$

$$q_y = -K_2\frac{\partial\theta}{\partial y} \qquad (3\text{-}14)$$

$$q_z = -K_3\frac{\partial\theta}{\partial z}$$

In (3-14), K_1, K_2 and K_3 are conductivities in the x, y and z directions, respectively. In case the conductivity is the same in the xy plane ($K_1 = K_2$) and different in the z direction, a case seen very often in sedimentary rocks, the heat flow could be expressed as:

$$q_x = q_y = -K_1\frac{\partial\theta}{\partial x} = -K_1\frac{\partial\theta}{\partial y} \quad \left(\text{since } \frac{\partial\theta}{\partial x} = \frac{\partial\theta}{\partial y}\right) \qquad (3\text{-}15)$$

$$q_z = -K_3\frac{\partial\theta}{\partial z}$$

Let us now consider a few crystal systems. In the cubic system, the conductivity is independent of the direction and hence the heat flow could simply be expressed as:

$$q_x = -K\frac{\partial\theta}{\partial x}$$

$$q_y = -K\frac{\partial\theta}{\partial y} \qquad (3\text{-}16)$$

$$q_z = -K\frac{\partial\theta}{\partial z}$$

In case of the monoclinic system, all classes have either a diad axis, the property that a rotation of 180° about this axis brings the crystal into a position congruent with its original one, or a plane of reflection symmetry. When the z-axis is either the diad axis or is normal to the plane of reflection symmetry, (3-13) reduces to:

$$\begin{aligned} q_x &= -\left(K_{11}\frac{\partial\theta}{\partial x} + K_{12}\frac{\partial\theta}{\partial y}\right) \\ q_y &= -\left(K_{21}\frac{\partial\theta}{\partial x} + K_{22}\frac{\partial\theta}{\partial y}\right) \\ q_z &= -\left(K_{33}\frac{\partial\theta}{\partial z}\right) \end{aligned} \tag{3-17}$$

In case of an orthorhombic system, all classes have either two perpendicular diad axes, or a diad axis with a plane of symmetry through it. When one of the axes is taken along a diad axis, and another along the second diad axis, or in the plane of symmetry, (3-13) reduces to:

$$\begin{aligned} q_x &= -K_{11}\frac{\partial\theta}{\partial x} \\ q_y &= -K_{22}\frac{\partial\theta}{\partial y} \\ q_z &= -K_{33}\frac{\partial\theta}{\partial z} \end{aligned} \tag{3-18}$$

For the hexagonal, trigonal and tetragonal systems, if the z-axis is a tetrad, triad, or a hexad axis (rotations of 90°, 120° or 60°, respectively), the reduction of equation (3-13) would be identical and the reduced form would be:

$$\begin{aligned} q_x &= -\left(K_{11}\frac{\partial\theta}{\partial x} + K_{12}\frac{\partial\theta}{\partial y}\right) \\ q_y &= -\left(K_{21}\frac{\partial\theta}{\partial x} + K_{22}\frac{\partial\theta}{\partial y}\right) \\ q_z &= -\left(K_{33}\frac{\partial\theta}{\partial z}\right) \end{aligned} \tag{3-19}$$

Heat flow in the triclinic crystal systems would be the same as in (3-13).

Fourier's equation (3-11), together with the energy conservation law can be used to derive a differential equation describing the temperature field in a medium:

$$\rho c \frac{\partial \theta}{\partial t} = \nabla (K \nabla \theta) \tag{3-20}$$

where ρ is density, t is time and ∇ is the Nabla operator.

HEAT CONVECTION

Within fluids, the heat transfer takes place through a combination of molecular conduction and energy transportation created by the motion of fluid particles. This mode of heat transfer is known as *convection*. The energy transport is called *free convection* or *natural convection* when the motion is produced due to an external force. In fluids, the heat exchange rate by convection is much higher than the heat exchange rate in solids through conduction. This difference becomes still more prominent in geothermics since rocks have very low thermal conductivities compared to metals and other solids.

In problems dealing with the transmission of heat through the process of convection, the fluid under consideration is usually bounded on one or more sides by a solid. Let at any given time θ_s be the temperature of the solid at its boundary with the fluid and θ_∞ be the fluid temperature at a far-off yet unspecified point. In accordance with Newton's law of cooling, the amount of heat flowing would be proportional to the temperature difference and could be expressed as:

$$q = h (\theta_s - \theta_\infty) \tag{3-21}$$

where h is the heat transfer coefficient. The heat is transferred by convection and consequently the heat transfer coefficient depends, in general, upon the thermal boundary condition at the solid-fluid boundary. However, under many situations, h can be estimated satisfactorily when the fluid dynamics of the flow system is known.

When a fluid flows past a stationary solid, the fluid viscosity causes the molecules adjacent to the solid to adhere to the solid surface and consequently velocity increases from zero at the solid surface to U_∞ at a far-removed point at the free stream. The velocity U reaches U_∞ asymptotically in accordance with the scheme shown in Fig. 3-1A. For practical applications, it is customary to define a layer thickness, S, where U attains an arbitrary value (say = $0.9\ U_\infty$). When the viscosity is low, S is usually small compared to the dimensions of the *external flow* systems (e.g., a system in which the fluid medium extends to much larger dimensions compared to the solid boundary area under consideration such as the flow over turbine blades, aircraft wings, etc.). In *internal flow* systems (e.g., flow in pipes, and ducts) the boundary layers are thin compared with the inlet aperture for a limited

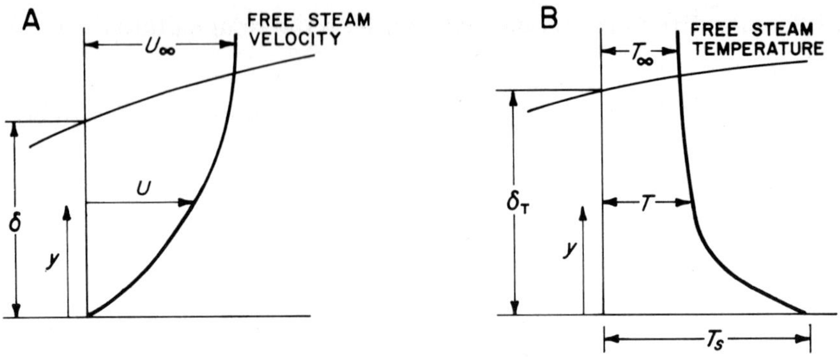

Fig. 3-1. Flow past a stationary wall. A. Velocity boundary layer. B. Temperature boundary layer.

distance only and as the distance from the entry point increases, the boundary layers merge, leaving no free stream region.

The temperature variations for a fluid flowing past a solid wall behave very similarly to the velocity variation discussed above. Fig. 3-1B shows schematically the temperature variations from the wall to the free stream temperature. Because of the fluid particles adhering to the wall, the heat from the wall could only be transmitted through molecular conduction and could be expressed as:

$$q = -K\left(\frac{\partial\theta}{\partial n}\right)_s \tag{3-22}$$

where $(\partial\theta/\partial n)_s$ is the fluid temperature gradient normal to the solid surface. When the temperature field is known, the heat transfer coefficient could be estimated from:

$$h = -K\left(\frac{(\partial\theta/\partial n)_s}{\theta_s - \theta_\infty}\right) \tag{3-23}$$

For external flow systems, θ_∞ is the uniform free-stream temperature and for the internal flow systems it represents the average of the bulk temperature of the fluid. Direct determinations of h from equation (3-23) are rare and it is mostly estimated from empirical or semi-empirical solutions of the fluid behaviour.

In the general three-dimensional case of flow involving interaction between temperature field and flow, it is necessary to know the velocity in the three directions, temperature, pressure, density, viscosity and thermal conductivity of the fluid for an adequate description of the fluid behavior. To reduce complexities, convection problems are often solved assuming a *steady state flow* where the dependent variables at a place do not change with time.

This simplification renders derivation of convection equations less difficult and is a fairly accurate description of a majority of thermal convection problems.

The following are the estimates of heat transfer coefficients under certain simple physical conditions.

At a surface of a square copper sheet of length of 50 cm with air flowing parallel to the surface with a velocity U between 0 and 25 m/s, the heat transfer coefficient h (cal/cm^2 s deg) is estimated to be (Nusselt and Jurges, 1922):

$$h = (1.71\ U^{0.78} + 1.28\ U^{-0.6U}) \times 10^{-4} \tag{3-24}$$

In case of a very rough surface and flow velocity of air exceeding 5 m/s, h would be:

$$h_{\mathrm{rough}} = 1.057\ (1.71\ U^{0.78} + 1.28\ U^{-0.6U}) \times 10^{-4} \tag{3-25}$$

In geothermics, one often needs to calculate convective heat transfer in mine shafts and galleries. Stoces and Cernik (1931) have estimated the coefficient of surface heat transfer between rocks and the air in galleries as:

$$h = (0.556 + 0.278\ U{\cdot}\alpha) \times 10^{-4} \tag{3-26}$$

where α varies from 5 to 10. Hiramatsu and Kokado (1958) have done some more sophisticated calculations taking into account the air resistivity at the wall of the galleries and the coefficient of the air friction in the air and found:

$$h = 3.47 \times 10^{-4}\ \frac{U_{\mathrm{m}} \nu K_{\mathrm{a}}}{V} \tag{3-27}$$

where U_{m} = air velocity averaged over the section of the gallery, ν = coefficient of friction in the air, K_{a} = thermal conductivity of the air, and V = viscosity of air.

Experimental measurements carried out in mines have shown that the above relations yield reasonable values for the coefficients of surface heat transfer.

HEAT RADIATION

In the previous sections, we have discussed the transfer of heat through conduction and convection, the two processes requiring presence of a medium. The means by which energy is transmitted between bodies without contact and in the absence of intervening medium is known as *radiation*. Transmission of energy through radio waves, visible light, X-rays, cosmic rays, etc., all belong to this category, having different frequencies in the spectrum of *electromagnetic radiation*. Here we are concerned with the type of radiation which is principally dependent on the temperature of the body, known as *thermal radiation* and belonging mostly to the infrared and to a small extent to the visible portion of the electromagnetic radiation spectrum.

The concept of *black body* is important in understanding the radiation of heat. According to the definition, a black body absorbs the entire amount of heat radiation falling on its surface. It also emits the maximum possible amount of heat at any given temperature, Q_b (cal/cm² s), which is expressed as follows for a unit area in a unit time:

$$Q_b = \sigma \theta^4 \tag{3-28}$$

where Q_b is the heat flow through radiation from the surface of a black body and σ is a constant known as the Stefan-Boltzmann constant, with a theoretical value of 5.67×10^{-8} W/m³ °K⁴ (= 1.38×10^{-12} cal/cm² s °K⁴). Since no material ideally fulfills the properties of absorption and emission of the theoretically defined black body, for practical purposes a new constant of emissivity, ϵ, is defined for real surfaces as:

$$\epsilon = Q/Q_b \tag{3-29}$$

Q being the radiant heat from the real surface. When measurements are made normal to the radiating surface, emissivity is called normal (ϵ_n) and the term ϵ generally implies total emissivity. In Table 3-2 the normal emissivity of a few solids is listed. From (3-28) and (3-29) the radiant heat flow from the surface of body with emissivity ϵ is:

$$Q = \epsilon \sigma \theta^4 \tag{3-30}$$

The heat radiated in a unit time from a unit area of an isothermal surface A

TABLE 3-2

Normal emissivity, ϵ_n, of a few solids

Material	θ (°C)	ϵ_n (cal/cm² s)
Gold, polished	130	0.018
Silver	20	0.020
Copper, polished	20	0.030
Copper, black oxidized	20	0.78
Aluminium paint	100	0.20—0.40
Iron, bright abrased	20	0.24
Iron, red rusted	20	0.61
Iron, heavily rusted	20	0.85
Zinc, grey oxidized	20	0.23—0.28
Lead, grey oxidized	20	0.28
Clay, fired	70	0.91
Porcelain	20	0.92—0.94
Glass	90	0.940
Paper	95	0.92
Wood	70	0.935
Tar paper	20	0.93

to another isothermal surface B with temperatures θ_A and θ_B and emissivities ϵ_A and ϵ_B would be:

$$Q_{AB} = b_{AB}\,(\theta_A^4 - \theta_B^4) \tag{3-31}$$

The constant b_{AB} depends upon the geometry of A and B and their configuration with respect to one another. When the two isothermal surfaces are parallel to each other and are separated by a distance which is small compared to their dimensions, b_{AB} becomes:

$$b_{AB} = \frac{\sigma}{\dfrac{1}{\epsilon_1} + \dfrac{1}{\epsilon_2} - 1} \tag{3-32}$$

When surface B encloses A, b_{AB} becomes:

$$b_{AB} = \frac{\sigma}{\dfrac{1}{\epsilon} + \dfrac{A}{B}\left(\dfrac{1}{\epsilon_2} - 1\right)} \tag{3-33}$$

In case a body with emissivity ϵ and temperature θ is surrounded by a black body at temperature θ_0, the approximate heat loss is given by:

$$Q = 4\,\theta_0^3\,\sigma\,\epsilon(\theta - \theta_0) \tag{3-34}$$

Relation (3-34) is used quite often in geothermics. In accordance with this relation, the temperature rise at the earth's surface assuming a terrestrial heat flow of 1.5 HFU and that all heat from the earth's surface is transferred through radiation and with $\epsilon = 0.9$ and $\sigma = 1.38$, would be:

$$\theta - \theta_0 = \frac{Q}{4\,\theta_0^3\,\sigma\,\epsilon} = 0.0137^\circ\text{C} \tag{3-35}$$

Since the heat loss at the earth's surface is not only through radiation, but also by convection, the real temperature rise observed at the earth's surface due to terrestrial heat flow is normally found to be less than 0.014°C.

TEMPERATURES WITHIN THE EARTH'S CRUST

Mathematical representation and analysis of precise thermal behavior within the earth is extremely difficult. Modelling is helpful in estimating gross properties and geothermal behavior under varying geological conditions. In the following we estimate the temperatures at shallow crustal depths under some very simplified assumptions.

Let us first consider a geothermally undisturbed section of crust, i.e., a section not disturbed by recent volcanic activity or any other tectonic or mechanical activity capable of disturbing the temperature field. Assuming that the thermal conductivity K and heat production H due to radioactivity, etc., are both dependent on depth, z, the temperature variation with depth could be expressed as:

$$\frac{d}{dz}\left[K(z)\frac{d\theta}{dz}\right] = H_0(z) \tag{3-36}$$

In the absence of a source of heat ($H_0 = 0$), the temperature θ_z at depth z could be expressed as:

$$\theta(z) = \theta_0 + q\int_0^z dz/K(z) \tag{3-37}$$

where q is terrestrial heat flow and θ_0 is surface temperature.

One of the geological settings often encountered is that of a layered structure (for example, a sedimentary rock country) in the crust. If the layer boundaries were to be horizontal and the thermal conductivity in each of these layers were to be constant, the temperature at the bottom of any layer m would be:

$$\theta(z_m) = \theta\left(\sum_{i=1}^{m} d_i\right) + \theta_0 + q\sum_{i=1}^{m}\frac{d_i}{K_i} \tag{3-38}$$

where d_i is the thickness and K_i the thermal conductivity of the ith layer. Suppose there were two layers of thicknesses d_1 and d_2 and conductivities K_1 and K_2 with horizontal interfaces and overlying a half space of conductivity K_3, then, in accordance with equation (3-38), the temperature at depth z would be:

$$\theta(z) = \theta_0 + q\frac{z}{K_1} \qquad \text{for } 0 \leqslant z \leqslant d_1$$

$$\theta(z) = \theta_0 + q\left(\frac{d_1}{K_1} + \frac{z-d_1}{K_2}\right) \qquad \text{for } d_1 \leqslant z \leqslant (d_1 + d_2) \tag{3-39}$$

$$\theta(z) = \theta_0 + q\left(\frac{d_1}{K_1} + \frac{d_2}{K_2} + \frac{z-d_2}{K_3}\right) \qquad \text{for } z \geqslant (d_1 + d_2)$$

In case the anomalous heat flow observed at the earth's surface could entirely be attributed to the heat production contrast of an anomalous body underground, the relation derived by Simmons (1967) could be applied. The relation, analogous to the one between gravity and magnetics, is simply:

$$\Delta q = \frac{gA}{2\pi G\rho} \tag{3-40}$$

where q = heat flow produced by the anomalous body, g = acceleration due to gravity, A = contrast in heat production, G = gravitational constant, and ρ = density.

Some relations obtained for some geometrically simple forms from (3-40) are as follows:

A. Sphere:

$$\Delta q = \frac{2}{3}\frac{Ar^3}{z^2}\left[1 + \left(\frac{x}{z}\right)^2\right]^{-3/2}$$

B. Vertical thin cylinder:

$$\Delta q = \frac{Ar^2}{2z}\left[1 + \left(\frac{x}{z}\right)^2\right]^{-1/2} \qquad (3\text{-}41)$$

C. Vertical sheet:

$$\Delta q = \frac{Ah}{2\pi}\left(\log\left[\left\{1 + \left(\frac{x}{z_2}\right)^{-2}\right\} \Big/ \left\{1 + \left(\frac{x}{z_1}\right)^{-2}\right\}\right]\right)$$

The various notations used in (3-41) are expressed in Fig. 3-2. Many more relations for simple two-dimensional as well as three-dimensional geothermal figures are given by Simmons (1967).

We shall now consider two examples of more realistic geological situations. In investigating geothermal fields over extended regions, the assumption of lateral homogeneity of thermal conductivity is mostly not valid. Changes in lateral conductivity effect the temperature field. When these changes could be approximated by simple geometrical boundaries, the corresponding temperature field could be estimated by application of mathematical approximation methods and use of modern computers.

In the finite difference method, the model is divided into a grid pattern. Let us consider the case when the model is divided into a square grid pattern with the length of a unit being Δd (Fig. 3-3). Using the Taylor series expansion, the derivatives occurring in the heat-diffusion equation are replaced by their finite differences. Let $\theta_{i,j,n}$ denote the temperature of the grid square (i,j) at a time $n\Delta t$ ($n = 0, 1, 2, \ldots$, and Δt is the time interval), and $\Delta\theta_{i,j,n+1}$ the increase in $\theta_{i,j,n}$ during next Δt time. $\Delta\theta_{i,j,n+1}$ can be calculated using the following relation (Hagiwara, 1966):

$$\Delta\theta_{i,j,n+1} = \frac{K\Delta t}{\rho c(\Delta d)^2}\left[\theta_{i-1,j,n} + \theta_{i+1,j,n} + \theta_{i,j+1,n} + \theta_{i,j-1,n} - 4\theta_{i,j,n}\right] \qquad (3\text{-}42)$$

where K = thermal conductivity, ρ = density, c = specific heat, and Δd = the length of the grid square.

The isotherms for an anticlinal structure (adopted from Mundry, 1966) are shown in Fig. 3-4. The middle deformed layer has the highest conductivity, causing the crowding of isothermal surfaces in the top and bottom layers of lower conductivity. The geometrical shape and thermal conductivity con-

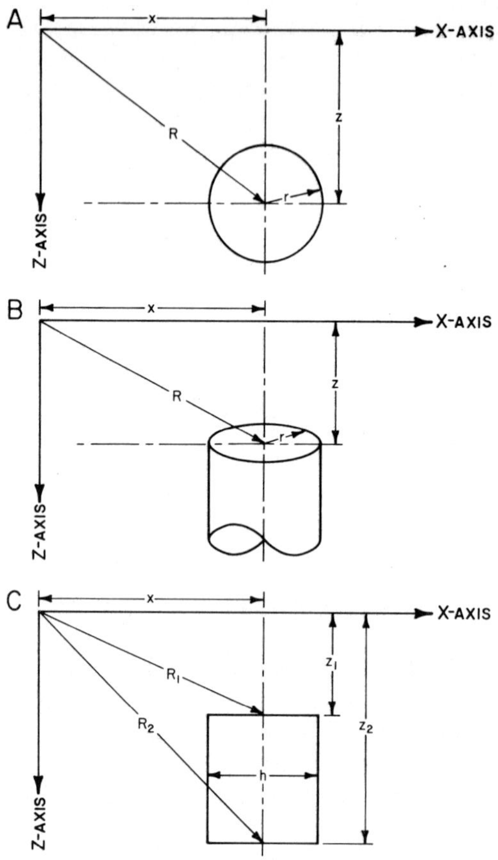

Fig. 3-2. Geometrical parameters used in relations (3-41) for (A) sphere, (B) vertical cylinder, and (C) vertical sheet.

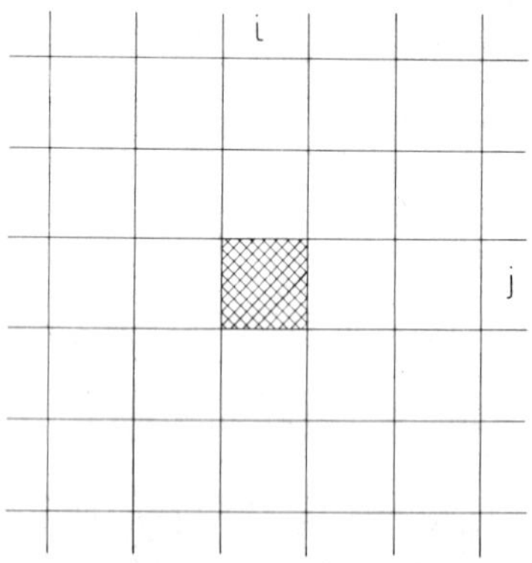

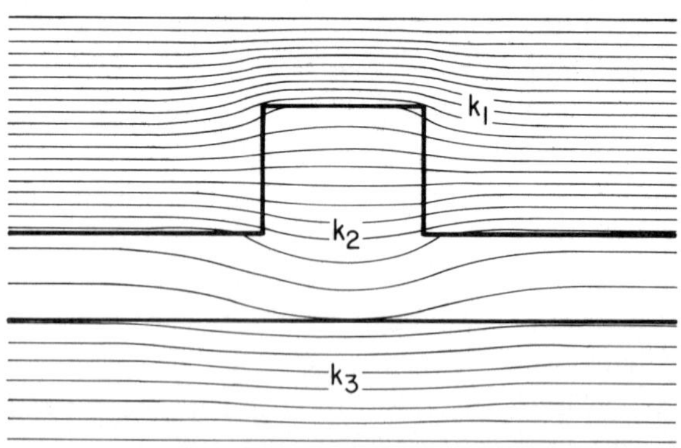

Fig. 3-3. Square grid pattern (see text).

Fig. 3-4. Isotherms for an anticlinal model structure with different thermal conductivities (after Mundry, 1966).

trast adopted in Fig. 3-4 is akin to a salt dome type of geological situation. It may be noted that the temperature distribution within the good conductor depends upon the conductivity contrast and the thickness of the overburden. Another example involving complex geology is presented in Fig. 3-5 (modi-

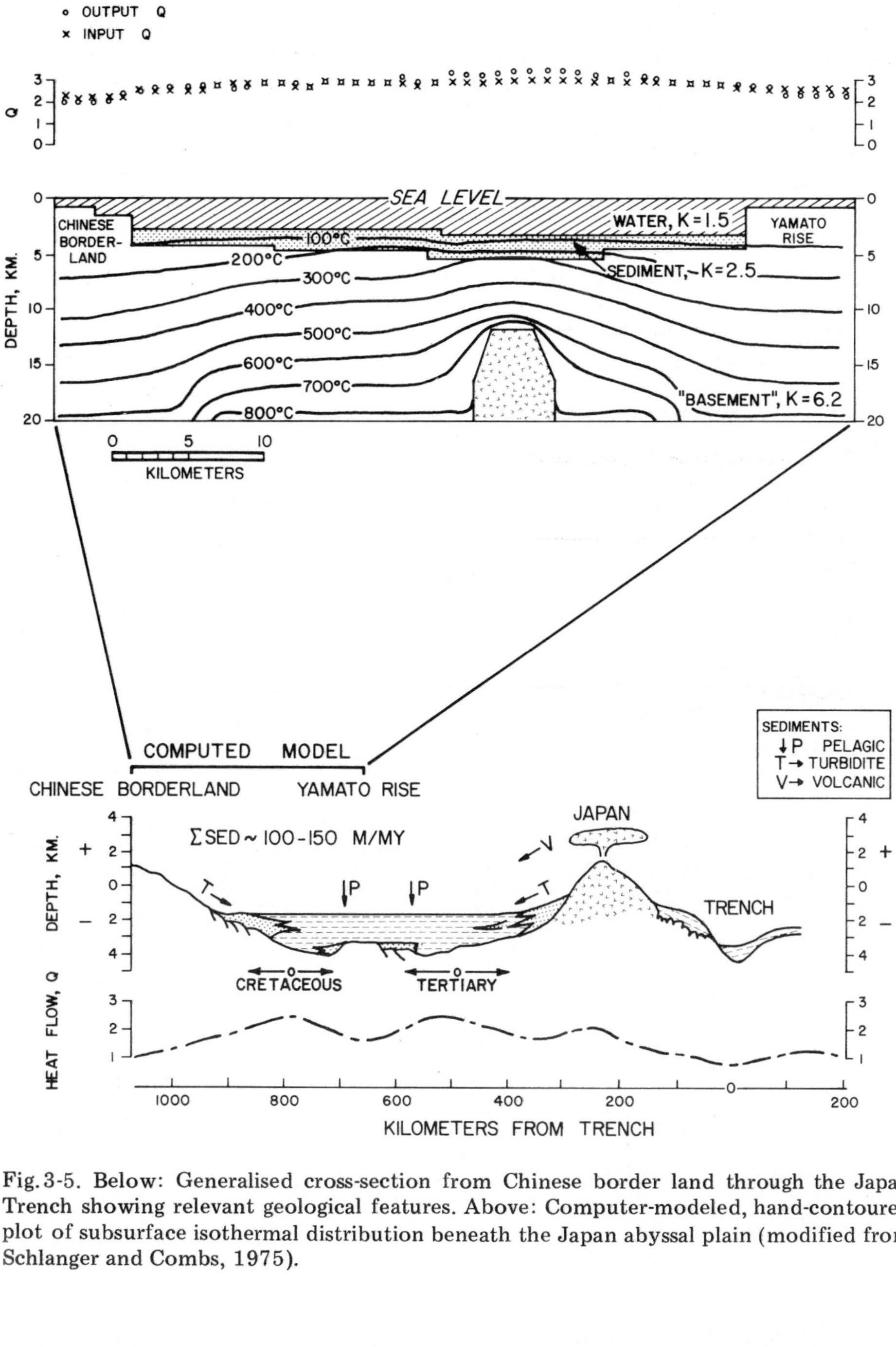

Fig. 3-5. Below: Generalised cross-section from Chinese border land through the Japan Trench showing relevant geological features. Above: Computer-modeled, hand-contoured plot of subsurface isothermal distribution beneath the Japan abyssal plain (modified from Schlanger and Combs, 1975).

fied from Schlanger and Combs, 1975). To investigate the temperature structure within the sedimentary prism in the Japan Basin, Schlanger and Combs (1975) used a finite difference approximation of the heat conduction equations and solved the resulting set of simultaneous equations by the method of successive over-relaxation. The thermal conductivity structure between the Chinese borderland and Yamato Rise was estimated from the isopach data and the geometry of the sedimentary prism. Finally, the subsurface isotherms were computed assuming a temperature of 4°C at the sediment-water interface and 800°C for the intrusive body so as to conform with the observed heat flow profile over the earth's surface.

Chapter 4

GEOTHERMAL SYSTEMS AND RESOURCES

INTRODUCTION

A geothermal resource can be simply defined as a reservoir from which heat can be extracted economically (costwise less expensive than or comparable with other conventional sources of energy such as hydroelectric power or fossil fuels) and utilized for generating electric power or any other suitable industrial, agricultural or domestic application.

Radioactive decay of unstable isotopes continuously generates heat within the earth. The heat content of the earth is estimated to be 3×10^{30} calories. However, under normal geothermal conditions, it is not economical to exploit this vast heat reservoir. The heat is lost from the earth's surface at an average rate of 1.5 HFU. It is believed that heat transfer below the lithosphere is mostly by convection and in the lithosphere by conduction. Rocks are relatively poor conductors of heat and below nine tenths of the earth's surface, the thermal gradients vary from about 10°C/km to about 60°C/km. Consequently, the temperatures encountered at depths of a few kilometers are in excess of the earth's surface temperatures by only an order of 100°C and power generators working at such small temperature differences have very low efficiency. With the improvement of drilling technology, it would become economical to drill deeper, say 5—10 km, and reach the required high temperatures of the order of 200—300°C. Nevertheless, it is not sufficient only to reach the hot rock, heat needs to be extracted by circulating fluids and for this process to be effective, there should exist an abundance of pore space and fissures for the fluids to permeate and circulate. At greater depths, the weight of the overlying rocks tends to close the pores and fissures, reducing the permeability considerably. Therefore, although hot rocks at depth exist everywhere, the current technological barriers, as well as unsuitable geological conditions make it uneconomical to extract geothermal energy in most places.

Under some geological situations, however, such as the plate boundaries and sometimes within them, heat may be locally transferred within a few kilometers of the earth's surface through the process of convection by magma or molten rocks. The magma has temperatures in the neighborhood of 1000°C and interacts with the near-surface rocks, causing surficial manifestations of geothermal activity such as the hot springs, geysers and fumaroles. Under certain suitable geological conditions, the heat becomes trapped, forming heat reservoirs. In such areas, after drilling to 100 m or so, temperatures of the order of 200—300°C are found and the regions could be suitable for harnessing the geothermal energy. In most of the geothermal fields under production, this sort of situation is being exploited.

TYPES OF GEOTHERMAL SYSTEMS

The essential requirements for a geothermal system to exist are (1) a large source of heat, (2) a reservoir to accumulate heat, and (3) a barrier to hold the accumulated heat. There is a suite of geological conditions that could result in a variety of geothermal systems. Consequently, all geothermal fields differ from one another. However, depending upon certain common characteristics, these can broadly be classified into the following categories: (1) vapor-dominated, (2) hot water, (3) geopressured, (4) hot dry rock, and (5) magma.

Difference of opinion continues to exist on many fine points concerning the details of the structure and mechanics of formation of these geothermal fields. Nevertheless, with the accumulation of new data, better understanding is being achieved. In the following, the above-mentioned categories are briefly discussed.

Vapor-dominated geothermal fields

Most of the presently exploited geothermal fields contain water at high pressures, and temperatures in excess of 100°C. When this water is brought to the earth's surface, the pressure is considerably reduced, generating large quantities of steam, and a mixture of saturated steam and water is produced. The ratio of steam to water varies from one site to another. Some of the best-known geothermal fields, such as Cerro Prieto (Mexico), Wairakei (New Zealand), Reykjavik (Iceland), Salton Sea (U.S.A.) and Otake (Japan), belong to this category. Since steam associated with water is produced in these fields, they are known as wet steam fields. There are a few other important geothermal fields such as Larderello (Italy) and The Geysers (U.S.A.) which produce superheated steam with no associated fluids. Fields of the latter type are known as dry steam fields.

The basic requirements of the vapor-dominated geothermal fields, whether of dry steam or of wet steam type, include adequate supplies of water in addition to the three prerequisites mentioned earlier. Fig. 4-1 is a schematic model representation of a vapor-dominated field.

Heat source. The fact that vapor-dominated geothermal fields are situated in regions of recent (Miocene-Quaternary times) volcanism, some of them being located on or close to volcanoes, has verified that magma is their source. Young, high-temperature (500—1000°C) magma intrusions to within depths of a few to several kilometers from the earth's surface allow the necessary heat to be accumulated in economical quantities. In hard compact rocks, faulting may provide a channel for the magma to reach the surface. Soft or plastic rocks, when present, can flow and block the fault space, causing the magma to spread at the contact between the soft and the hard rocks. Active volcanoes, fumaroles, hot springs and geysers are obvious surface manifestations of recent volcanic activity. In addition, certain geological environments,

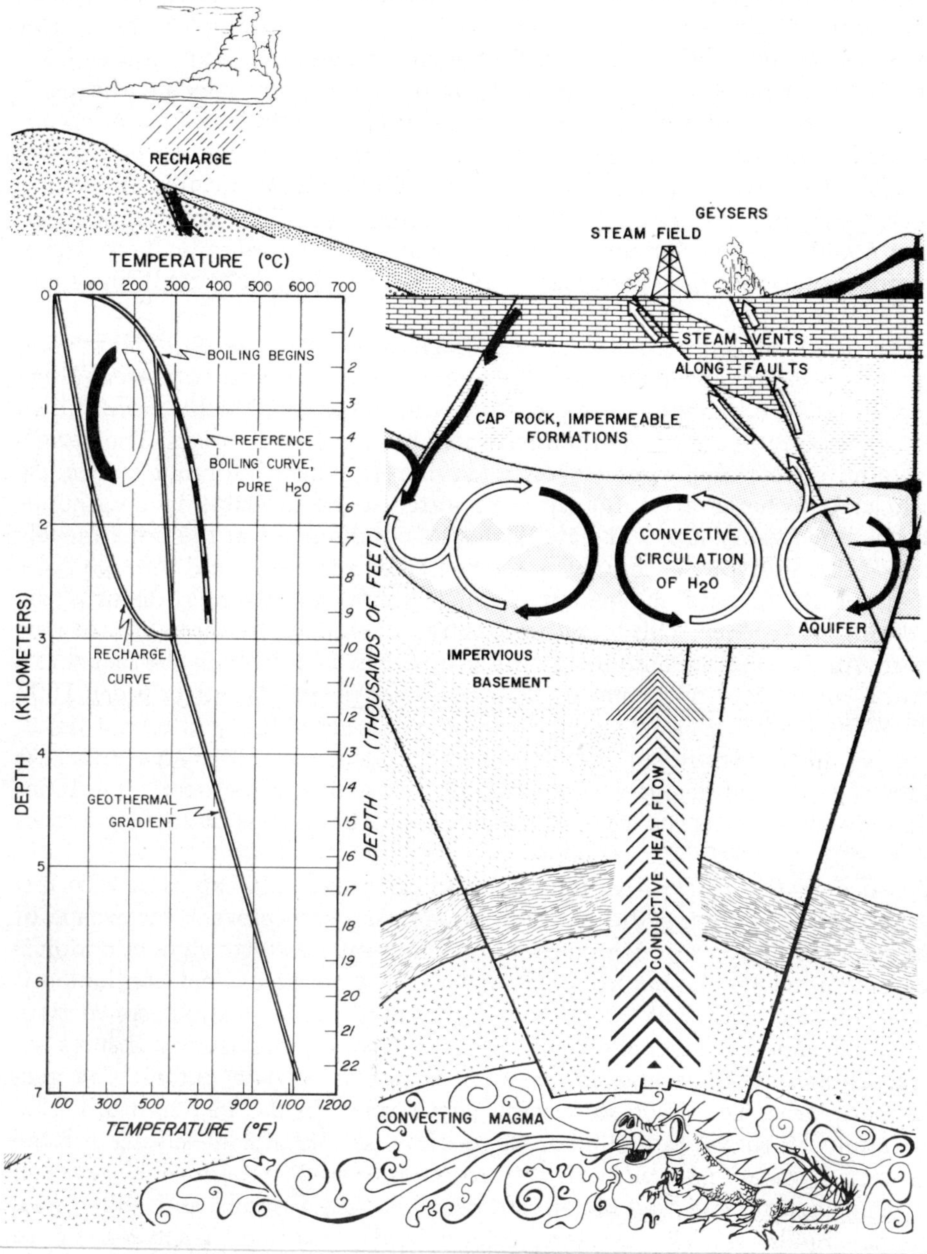

Fig. 4-1. Conceptual model of a steam geothermal system.

such as regions of Quaternary uplift and regions of Late Tertiary and Quaternary subsidence, are indicative of shallow magmatic intrusions.

Reservoir and water supply. In order to form a heat reservoir, the anomalous magmatic intrusion should encounter porous and permeable, water-filled rock strata. Within the reservoir, convection currents of hot water and/or steam are set up, providing a good heat exchange, and the temperature difference between the top and bottom of the reservoir is not very significant. A variety of rocks have been found to constitute the lithology of a reservoir. At Larderello (Italy) it is fractured limestone and dolomite, at The Geysers (U.S.A.) it is fissured greywacke, at Wairakei (New Zealand) it is pumiceous breccia and tuff, and at Cerro Prieto (Mexico) it is deltaic sands. Good reservoirs could also be formed at unconformities and formation boundaries, provided that they are permeable and have good hydraulic continuity and water supply.

The origin of geothermal fluids has been debated in the past and in addition to meteoric, magmatic and juvenile origins have been suggested. However, recently conducted isotopic studies in geothermal fields have shown that at least 90% of the geothermal water has a meteoric origin. The permeable aquifers forming the reservoir must therefore have hydraulic continuity with large recharge areas for the rain water to be available in continuous supply. The freshly supplied water is heated conductively at the impermeable base of the reservoir. Withdrawing of the heated reservoir fluid through boreholes, or its upward movement through vents and fissures, disturbs the hydrological balance. This is restored, fully or partially, by the inflow of new water. An idea of the amount of the inflow can be had from the fact that a natural steam field operating a 100-MW power plant lets out between 1000 and 2000 metric tons of water every hour. Some of the geothermal fields, such as the Larderello in Italy, have easily identifiable recharge areas. At Larderello, the permeable reservoir terrain, consisting of Mesozoic limestones and dolomites, outcrops thereby providing an easy access to surficial water.

Cap rock — the barrier. An impermeable cap rock, or a cap rock with low permeability, overlying the reservoir, is necessary to prevent the escape of hot reservoir fluids through convection. The heat loss through conduction is not prevented by the cap rock. However, the amount of heat conducted is much smaller than that which could be lost through possible convection. Since volcanism is associated with tectonic movements causing fissures, an ideal unfissured impermeable cap rock is nowhere to be found. The geochemical processes associated with geothermal fields, i.e., hydrothermal alteration of rocks and mineral deposition, are helpful in sealing off the fissures. Typical examples of cap rocks rendered impermeable through chemical action and deposition are seen at The Geysers and Otake geothermal fields. At The Geysers, calcite- and silica-filled fractures, up to 1 inch wide, are commonly seen. Evidence of hydrothermal alteration is presented by the bleaching of greywacke as well as by the absence of vegetation in patches. The geochemical and hydrothermal processes are complicated and vary from place to place.

At many other steam-producing fields, original impervious rocks constitute the cap rock. Examples are the lacustrine Huka Formation at Wairakei (New Zealand), the deltaic clay at Cerro Prieto (Mexico) and Salton Sea (California), and the Flysch Formation at Larderello (Italy).

Hot water geothermal systems

In hot water geothermal fields, water convection currents carry the heat from the deep source to the shallow reservoir. The bottom of the convective cell may be heated through conduction from hot rocks. The geology of hot water geothermal fields is quite similar to that of an ordinary groundwater system. They differ from the earlier-discussed vapor-dominated geothermal fields in the fact that the hot water geothermal fields are characterized by liquid water being the continuous pressure-controlling fluid phase. Typically, the temperature of hot water reservoirs varies from 60 to 100°C and they occur at depths ranging from 1500 to 3000 m. Fig. 4-2 is a schematic representation of a hot water field and the reference curve represents the variation of the boiling point of the pure water with depth. As shown in Fig. 4-2, a hot water geothermal field could develop in the absence of a cap rock, if the thermal gradients and the depth of the aquifer are adequate to maintain a convective circulation. When the cap rock is absent, the temperatures in the upper part of the reservoir cannot exceed the boiling point at atmospheric pressure, since with the convective rise, the water loses pressure and also becomes mixed with the cool groundwaters.

Depending upon the temperature, chemistry and the structure of the reservoir, hot water systems have been classified into several subtypes. These have been discussed in some detail by White (1974). The following is a brief description of various subtypes:

(1) Systems characterized by low to moderate temperatures (say, 50–150°C) and producing water with a chemical composition similar to the regional surface and shallow groundwaters.

(2) Systems characterized by the presence of partly non-meteoric water. Such systems usually occur in deep sedimentary basins.

(3) Systems characterized by the presence of brine of very high salinity. The chemistry can vary considerably from one field to another. The Salton Sea (California) and the Red Sea brine pools belong to this subtype and have very differing bulk chemistry of the sediments and the associated rocks, probably attributable to the difference in brine composition.

(4) Systems characterized by the presence of natural cap rocks. Geothermal fields at Cerro Prieto (Mexico) and Salton Sea (California) have cap rocks constituted by fine-grained, low-permeability sediments.

(5) Systems characterized by the creation of their own self-sealing cap rocks. As explained earlier in discussing vapor-dominated geothermal sys-

tems, these cap rocks are formed through chemical alteration and deposition of sediments near the surface where the temperature decreases suddenly. Wairakei (New Zealand) and Yellowstone Park (Wyoming) are typical examples.

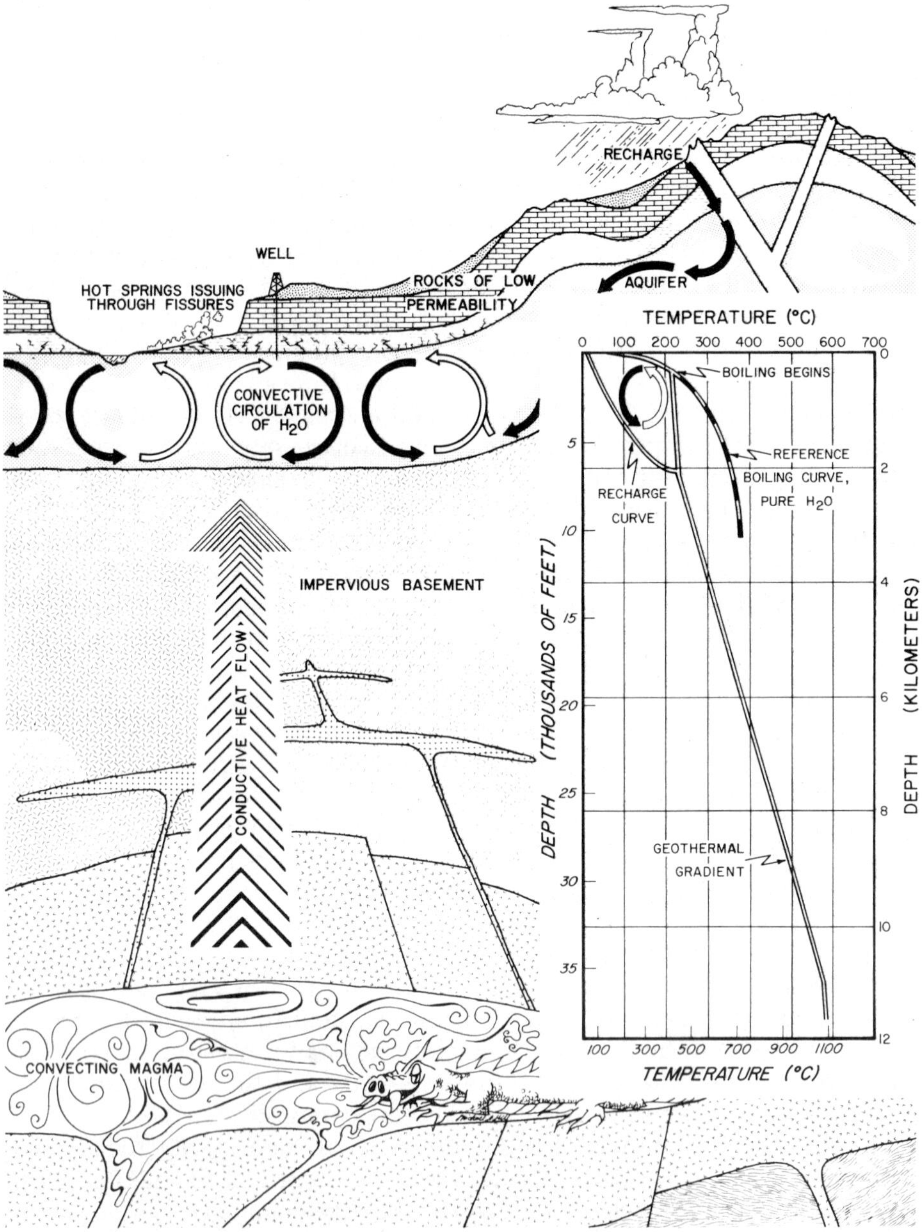

Fig. 4-2. Conceptual model of a hot water geothermal system.

Geopressured geothermal resources

Recently geopressured geothermal resources have drawn much attention as possibly very potential sources of geothermal energy. Studies on the Gulf of Mexico geosyncline, for which much geological information exists because of petroleum exploration, have been extremely useful in understanding the mechanics and geology of the formation of the geopressured geothermal systems. Large-scale subsidence is presently occurring in modern geosynclines involving downward transformation of enormous amounts of sediments to depths of 5—10 km in relatively short geological time intervals. As evidenced in the Gulf of Mexico, normal faulting is the most important structural feature and faults involving throws of up to 1 km are not uncommon. This process of subsidence subjects the poorly consolidated sediments to intense heat and enormous pressure. As pointed out by Jones (1970), the process of geosynclinal subsidence and rapid sedimentary deposition is analogous to igneous intrusion in an opposite sense: the motion is downwards, the intruding material is cold and the rocks intruded are hot. If the sediments undergoing subsidence are constituted of clay of a swelling variety, endothermic diagenesis of clay minerals can consume much heat and thereby reduce the heat flow to the earth's surface. This phenomenon is very noticeable in the Gulf of Mexico Basin where sediments are mostly of the swelling variety of clay and consequently the geothermal gradient map of the area shows it as a negative geothermal area (Fig. 4-3). Based upon measurements in holes less than 2 km deep, this map shows a decrease in geothermal gradient along the axis of the geosyncline. Obviously, being dependent on shallow heat flow measurements, the map does not indicate (or conceals)

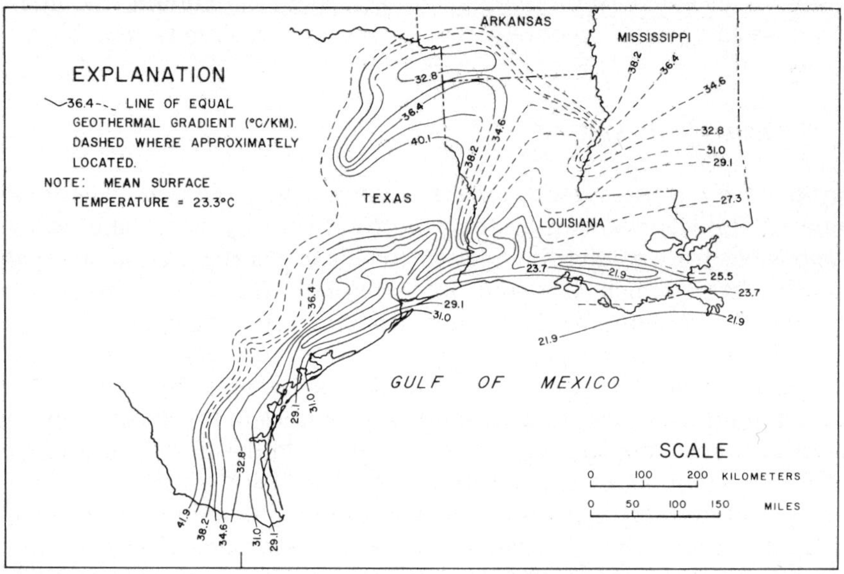

Fig. 4-3. Geothermal gradients in the Gulf Coast of the United States (from Jones, 1970). For details, see text.

the true magnitude of the existing geothermal resource potential in the northern Gulf of Mexico Basin. It has been observed that the geothermal gradient in the hydropressured zones at depths ranging from 1 to 2 km varies from 20 to 40°C/km, and it undergoes a very sharp increase in the geopressured zone; being two to three times greater than that existing in the overlying hydropressured zone.

The deeply buried clays in the geosynclines undergo thermal metamorphism, releasing petroleum hydrocarbons and geothermal waters, the former being solute and the latter solvent, and a solution is formed. The strength of the load-bearing clay bed is drastically reduced by the process of metamorphism, resulting in the expulsion of the pore fluids to share the overburden load. Since the confining pressure decreases most rapidly in the upward direction, and the pressure applied to a confined fluid is transmitted equally in all directions, the expelled pore fluid moves out of the clay bed and moves upwards through sand beds and along the fault planes. The expelled solution of hydrocarbons and water, as it moves upward, encounters much lower pressures and temperatures. Finally, a stage is reached when the temperatures and the pressures are so low that the solute hydrocarbons can no longer remain in the solvent water. Fluid hydrocarbons and geothermal waters then appear as separate phases, and the hydrocarbons being lighter than the water seek higher points in the regional flow system and structural traps. Detailed geological mapping of structures and knowledge of sedimentary history, when combined with the isothermal, isosaline and isopressure maps are helpful tools in prospecting fluid hydrocarbon and geothermal regimes.

To date, potential geopressured geothermal fields have been discovered in the Gulf of Mexico coast area only. However, with the accumulation of more data and the advancement of drilling technology, there is no doubt that many more commercially viable geopressured geothermal reservoirs will be discovered.

Hot dry rock geothermal systems

So far we have discussed geothermal resources associated with one or another type of fluid convection system; the fluid (mostly steam and water) being responsible for carrying the heat to the earth's surface or to the shallower depths from which it could be exploited. However, there exists another category of geothermal resource where thermal anomalies exist at shallow depths within the earth's crust without any fluid availability to store or transport the heat. These resources are designated hot dry rock. After having described the hot dry rock geothermal resources in the above manner, three questions arise immediately: "how deep?", "how hot?", and "how dry?" Different depth and temperature ranges as well as definitions of "dryness" can be found in the literature. Here we adopt the definition given by the U.S. Energy Research and Development Administration (ERDA) in their Report No. 1 of the Hot Dry Rock Assessment Panel to answer these questions concerning the hot dry rock geothermal resource. Accordingly, it is "heat stored in rocks within 10 km of the earth's surface from which the energy cannot be economically produced by natural hot water or steam". Additionally, they cannot produce an economical volume of hot water or steam and their temperature is less than about 650°C (so as to exclude molten lava or magmas). Quite understandably, hot dry rocks may be associated at some locations and/or at some depths with hydrothermal resources and/or magma.

Depending upon the cause of the hot dry rock geothermal resources, they can broadly be classified into three categories. These are: (1) igneous related: heat being transferred from magma or stored in dry rocks surrounding the magma bodies; (2) upper mantle related: the heat being conducted to shallow crust overlying an unusually hot upper mantle, causing the anomaly; and (3) local: heat being stored locally either due to the presence of a high concentration of radioactive minerals or due to large-scale faulting and/or fracturing.

Considering the above classification of hot dry rocks based upon the different possible sources, the obvious places to look for them are the regions characterized by: (1) recent volcanism, (2) high heat flow, and (3) localized radiometric heat sources. All magma chambers have a surrounding hot rock environment of varying size. Similarly, geothermal resource areas characterized by hydrothermal convective systems will, at some depth, be underlain by hot dry rock. Also, high heat flow areas, where the basement geology is favorable for low water contents, are possible hot dry rock resource areas.

Magma

Magma, the naturally occurring molten rock material, is a hot viscous liquid which retains fluidity till solidification. It may contain gases and particles of solid materials such as crystals or fragments of solid rocks. However, the mobility of magma is not much effected until the content of solid material is too large. Typically, magma crystallizes to form igneous rocks at temperatures varying, depending upon the composition and pressure, from 600 to 1400°C. At its site of generation, magma is lighter than the surrounding material, and consequently it rises as long as the density contrast between magma and surrounding cooler rocks continues. Eventually, magma either solidifies or forms reservoirs at some depth from the earth's surface, or it erupts. Plate boundaries are the most common sites of volcanic eruptions (Fig. 2-7). It has been estimated that the average rate of production of magma at Kilauea Volcano, Hawaii, during the period of study (1952—1971) has been about 10^8 m^3 per year (Swanson, 1972). Similar estimates have been made for the Columbia Plateau (Baksi and Watkins, 1973) and elsewhere. The heat energy available from such sources, if harvested, is an order of magnitude higher than the total annual energy requirements of the largest energy-consuming country in the world: the U.S. energy consumption for the year 1972 was estimated to be 0.7×10^{20} calories. However, up to the present, the necessary technology has not been developed to recover heat energy from magma.

Chapter 5

EXPLORATION TECHNIQUES

The contemporary economic and technological state of affairs and its foreseeable development, restricts extraction of geothermal energy to the upper few kilometers of the earth's crust. The deepest geothermal wells, to date, are less than 5 km deep. The development of drilling technology and increase in the cost of other energy resources will certainly permit of drilling deeper for geothermal resources in the near future. However, it is unlikely to exceed 10 km in depth. The recently developed plate tectonics hypothesis provides useful clues as to where to look for geothermal resources. Once a geothermal region is identified, the next step is to use different available exploration techniques to localize potential geothermal resource areas and identify suitable drilling targets for production. It is necessary to estimate the temperature, volume and permeability at depth as well as to predict whether the well will produce dry steam, wet steam or just hot water. It is also desirable to estimate the chemical composition of the fluid to be produced. To obtain this varied information, it is necessary to apply a suite of exploration techniques. Important among these are: (1) geological and hydrological techniques, (2) geochemical techniques, (3) geophysical techniques, and (4) airborne surveys.

Before undertaking any of these detailed surveys, it is necessary to undertake a detailed search and collection of relevant existing literature. For many areas, useful geological, hydrological, geochemical, geophysical, topographical and meteorological information already exists, and this should be fully utilized before undertaking new exploration. It is very likely that the literature survey could provide useful information regarding the suitability or otherwise of the various exploration techniques discussed in the following.

GEOLOGICAL AND HYDROLOGICAL TECHNIQUES

General background

The first step in selecting a region for preliminary reconnaissance is to look for obvious manifestations of geothermal resources such as thermal springs, fumaroles, geysers, etc. A detailed list of all these manifestations, including a description of the locality, physical environment, temperature and amount of the fluid discharged, should be compiled. Many countries have long-established geological surveys and such information is often well documented. The location of these geothermal manifestations should next be studied on the most detailed geological maps available. This would reveal whether these manifestations are associated with a particular rock type or structural zone. If they tend to be associated with a particular zone or district, then that zone is likely to be an area of positive geothermal anomaly. It is also likely that there are many more geothermal springs in the region which have not been reported. A field reconnaissance is desirable, with particular attention being given to areas where existing information has indicated potential geothermal resource, and extended to all hydrothermal areas. It is necessary to carry out geochemical analysis of the water from springs.

Geothermal resources, on the basis of their origin, can broadly be classified into two main categories and various subcategories (Muffler, 1976a):

(1) Geothermal resources associated with igneous intrusions in the upper crust. These can be further divided into: (a) magma, (b) hot dry rock, and (c) convective hydrothermal systems.

(2) Geothermal regimes not related with young igneous intrusions, further classified into (a) resources associated with deep circulation of meteoric water, (b) high-porosity environment resources at hydrostatic pressure, and (c) high-porosity environments at pressures greater than hydrostatic.

In the second chapter, the concept of the plate tectonic hypothesis has been introduced. This hypothesis is extremely useful in geographically identifying broad areas in which the two primary categories of geothermal resources are found. Fig. 2-7 shows the present-day configuration of the lithospheric plates and the location of the major geothermal fields. Plate boundaries are characterized by earthquake occurrences. Generation of magma takes place at the spreading ridges, and at the trenches where plates are subducted, as well as at intraplate melting anomalies. Therefore, it is to be expected that geothermal resources related with the intrusion of magma in the shallow crust would occur primarily close to the spreading ridges, subduction zones and intraplate hot spots like Hawaii. However, geothermal resources of conductive crustal regime could occur throughout.

Geothermal resources associated with igneous intrusions in the upper crust

The three varieties of geothermal resource associated with the intrusion of igneous rocks, i.e., magma, hot dry rock and convective hydrothermal reservoir, require emplacement of a body of magma in the upper crust. It is necessary that a portion of magma be prevented from erupting by some suitable barrier. When magma from the lower crust or the upper mantle quickly reaches the earth's surface, it looses all its heat to the atmosphere without producing a significant geothermal resource. Once a body of magma is confined at some depth from the earth's surface, a variety of conductive and convective processes dissipate its heat, giving rise to geothermal resources.

As discussed earlier, magma is generated at mid-oceanic ridges and hence they are potential geothermal resource targets. However, practical problems render them less attractive and to date energy is harnessed only in Iceland, where the Mid-Atlantic Ridge is above sea level. When spreading ridges impinge on a continent, partial fusion of crustal material forms silicic magma. Therefore, silicic intrusions and extrusions are most commonly observed where spreading ridges impinge on continents, compared to mid-oceanic sites (Robinson et al., 1975).

At the zones of subduction, material is thrust down into the lower crust and upper mantle. The composition of this material varies from siliceous sediments to ultramafic rocks. Its partial melting produces a variety of silicic and intermediate rocks. The material properties of the subducting plates, i.e., whether it is an oceanic plate—oceanic plate subduction, or an oceanic plate—continental plate subduction, or a continental plate—continental plate contact, govern the composition of the rocks produced.

Intra-continental melting anomalies have been postulated by Morgan (1972) to lie above narrow plumes (less than 150 km in diameter) that originate deep in the earth's mantle. Propagating fractures and shear melting have also been suggested as being responsible for melting anomalies (McDougall, 1971; Shaw and Jackson, 1973). The best-known example of such anomalies is beneath Hawaii where large volumes of basaltic magma have erupted from depths of at least 60 km (Dalrymple et al., 1973).

Depending upon the existing geological and geohydrological conditions, the three major categories of geothermal resources develop consequent on the intrusion of large volumes of molten rocks into the shallow crust. The magma itself is a potential resource. According to the estimates of Muffler (1976a), a cubic kilometer of granitic magma at 800°C contains 3×10^{18} joules of heat, which is equivalent to the heat content of 480 million barrels of crude oil. However, the technological problems in harnessing geothermal energy from magma bodies are enormous. After intrusion in the crust, magma looses heat through conduction and convection. When the permeabilities of the host rock and the intrusion are low, a hot dry rock geothermal resource is developed. One cubic kilometer of hot granite at 400°C contains 1×10^{18} joules of heat. The technological problems involved in utilizing the heat of hot dry impermeable rocks are substantial. Proposals for hydraulic fracturing of the rocks and circulating pressurized water in closed loops are being considered. When an igneous intrusion is emplaced in rocks with adequate fractures or intergranular permeability, the cooling of the igneous intrusion is considerably affected by movement of water. Under favourable conditions of hydrological continuity, meteoric water enters through permeable rocks, comes in contact with the hot intrusion, becomes heated and moves upward, setting up strong convective currents. The heat from the intrusion is thus either convected to the earth's surface or it forms a hydrothermal reservoir at some intermediate depth. Depending upon the local hydrological situation, this could be a vapor-dominated or a hot water reservoir.

From the above it may be concluded that any one crustal geothermal anomaly, caused by igneous intrusion, can support a variety of geothermal resources simultaneously and/or sequentially. The potential of an igneous system is directly proportional to the size of the anomaly and indirectly proportional to its age. In Fig. 5-1, empirical relations derived by Smith and Shaw (1975) are shown. They have considered the age-volume data of 54 volcanic systems in the U.S.A. and plotted the approximate present position of each system in relation to its probable solidification state and to the 300°C isotherm. Smith and Shaw (1975) have drawn pairs of lines to represent a spectrum of cooling models that identify igneous systems which are approaching ambient temperatures (points above lines *5* and *6*), systems that may now be approaching the post-magmatic stage (points between lines *3* and *4*), and systems which may still have magma chambers with large portions of molten magma (points below lines *1* and *2*). The pair of lines covers geometric shapes varying from cubes to slabs. Lines *1* and *2* are drawn assuming that cooling takes place by internal convection till solidification takes place. Lines *3* and *4* assume that cooling is due to conduction only, both inside and outside the magma chamber. Lines *5* and *6* represent an estimate of the time required before the central temperature of the solidified pluton has fallen to about 300°C. Points lying above line *6* are considered to be of little geothermal potential. Smith and Shaw (1973, 1975) have used these plots for reconnaissance evaluation of silicic volcanic areas.

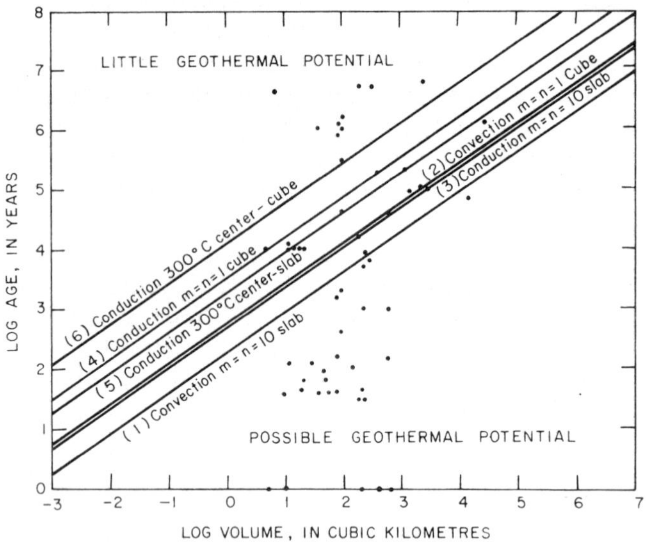

Fig. 5-1. Graph showing theoretical cooling time versus volume of magma bodies. See text for explanation of lines *1*–*6* (from Smith and Shaw, 1975).

Geothermal regimes not related to young intrusions

Resources associated with deep circulation of meteoric water. Warm water springs are sometimes found in regions characterized by normal, or even low, geothermal gradients. Moreover, these regions are not found to be associated with Cenozoic volcanism. Geothermometry, based on geochemical analysis, does not reveal elevated temperatures at shallow depths. Such warm springs appear to be associated with deep circulation of meteoric water. Warm springs of the Basin and Range Province of the western United States is one such example. The Basin and Range Province area is characterized by extensive normal faulting and tectonic extension (Thompson and Burke, 1973). The hot springs are mostly located along faults. Another well-known example is that of the Larderello field in Italy (Marinelli, 1964), where a widely spread impermeable cap rock overlies a folded sedimentary series with a highly permeable member. The basic requirement for this kind of geothermal reservoir to exist seems to be the presence of hydraulic continuity to great depths and some sort of a barrier to form a reservoir.

Resources associated with a high-porosity environment at hydrostatic pressure. Deep basins filled with sedimentary rocks of high porosity and permeability and occurring in regions of high geothermal gradients are favored sites of geothermal resources at hydrostatic pressures. High temperatures, approaching 200°C, can be expected at shallow depths. The Hungarian Basin, where porous Cenozoic sediments, up to 3 km thick, occur in a region of high-temperature gradients of 50—70°C/km, is one such example. A water reservoir comprising over 4000 km^3 of water at temperatures varying from 60 to 200°C is estimated to exist in these porous rocks (Boldizsár, 1970). Geothermal energy from similar resources is being harnessed in the U.S.S.R. and France. The two basic requirements for this kind of geothermal resource to occur are the existence of high regional heat flow and a sedimentary basin with a thick pile of porous and permeable rocks. Plate margins are favored to satisfy these requirements (Muffler, 1976a).

Resources associated with a high-porosity environment at pressures in excess of hydrostatic. Zones characterized by pressures greatly in excess of the hydrostatic pressure, have been revealed by drilling carried out in sedimentary basins. These zones are called geopressured zones and are associated with high temperatures. Most of the information on geopressured geothermal resources has accrued from investigations carried out in the Gulf Coast of the United States and summarized by Jones (1970) and Papadopulos et al. (1975). This has been discussed in Chapter 4.

As explained by Lewis and Rose (1970), these extensive geopressured reservoirs are bounded laterally by faults and they occur in sands underlying low-permeability shales. Thermal insulation is provided by the overlying clays filled with static water. Whether the regional heat flow in the Gulf Coast is high or not, is not known. However, the relatively shallow depth of the Moho and the discovery of basalts of unknown age suggest elevated regional heat flow (Muffler, 1976a).

The basic requirement for geopressured geothermal reservoirs to occur seems to be the rapid rate of sedimentation in shallow basins characterised by high regional heat flow.

No geothermal energy has been yet harnessed from geopressured geothermal resources. With technological and other developments, and as and when it becomes possible to harness these resources, they will provide an abundant source of energy.

Choosing a prospect area

The above-mentioned geological and geohydrological considerations are useful in broadly selecting a promising region for geothermal prospecting. The next step is to select specific areas for detailed prospecting. In areas where geothermal manifestations are not well documented, field surveys should be undertaken to locate them and to collect information on the type of manifestation (i.e., spring, well and fumarole, seepage, CO_2 vent, etc.). Such surveys can be very time-consuming. If adequate funds are available, remote-sensing surveys, discussed later, to locate thermal manifestations should be undertaken.

As summarized by McNitt (1973), surface manifestations may not reflect conditions at depth. This is particularly true when an overlying non-geothermal groundwater horizon masks the geothermal system. For this reason, it is not always possible to use surface heat discharge as a first approximation to the size of the system and its capacity to produce usable energy. For example, the Larderello fields in Italy, The Geysers and the Salton Sea fields in the United States and many other geothermal fields have meager surface heat discharge compared to the amount of energy obtained by drilling. At places, productive ground is found to extend many kilometers beyond a small surficial manifestation. It is particularly difficult to estimate the spread and capacity of vapor-dominated geothermal systems from surface indications. Such systems require a cap (as discussed in Chapter 4) for building up of the high temperatures and pressures in the trapped vapor. This cap prevents excessive surface indications. However, for hot water geothermal systems, a lithological or hydrological cap is not necessary. Consequently, for such systems, surficial thermal manifestations can be used to approximate their size. The Yellowstone geothermal field in the United States and Wairakei in New Zealand are excellent examples of this kind. At places impermeable lithological horizons tend to deflect the upflowing hot water laterally for several kilometers. This results in surficial thermal manifestations being remote from the center of the subsurface geothermal reservoir. These lithological barriers are commonly caused by the fine sediments of streams or lakes flowing at the top of the geothermal systems. These sediments are deposited along the fault angles and cause the upcoming hot water to move up dip beneath them for long distances before surfacing. As mentioned by McNitt (1973), at the Kizildere field in Turkey, this kind of lithological situation causes deflection of the upcoming thermal fluid from the center of the subsurface geothermal reservoir by about 5 km. This kind of lateral deflection should be suspected in the regions where the geothermal springs occur at the contact of alluvial beds with the bed rocks rather than occurring in fault or fracture zones.

Certain hydrological conditions also considerably alter the possible surficial manifestation of geothermal reservoirs. When the top of a geothermal system comes in contact with a deep underground body of water, where the water is stagnant or slowly moving, and the heat from the geothermal system is adequate to cause the groundwater to boil, patches of gently steaming ground would appear. These patches would give no indication of the high pressures that may exist in the geothermal systems. Similarly, when the top of a geothermal system is intersected by a large cold water aquifer, the heat will be swept down gradient (McNitt, 1973). Therefore, either no evidence of the geothermal system will appear on the surface, or large-volume warm springs may appear at large distances from the center of the source of heat.

In conclusion, as summarized by Healy (1970), the following stages of investigation are suggested.

(1) Review of the existing data on thermal springs and preparation of a map showing their relation to regional geology. This could reveal areas where more attention is required.

(2) Field reconnaissance of the thermal springs, sampling of the water for chemical analysis, and study of aerial photographs when available.

(3) Detailed study of the characteristics of geothermal areas. These characteristics include the size of the area, the presence or absence of shallow and deep aquifers, subsurface temperatures, and potential for development as steam or hot water field, etc.

(4) Investigation of the hydrothermal areas and their geological characteristics to assess their relative importance from genetic, structural and regional points of view.

(5) On the basis of the above studies a decision is to be made whether production of geothermal energy appears possible or not. If there are positive indications, several areas should be demarcated in order of preference for further investigations.

GEOCHEMICAL TECHNIQUES

Over the last two decades, geochemical methods have been increasingly applied in geothermal resource development. They are relatively much less expensive as compared to geophysical surveys and subsurface investigations by drilling and this has made them more popular. Geochemistry plays an important role in preliminary prospecting and serves as a useful guide for decision making as to where drilling should be undertaken. During the development of a geothermal field, geochemical analysis of geothermal fluids provides information on flow patterns of subsurface fluids. Geothermal fluid chemistry is widely applied in exploration. Recent geochemical exploration carried out in various countries has resulted in the development of new methods for estimating subsurface temperatures.

Here we shall briefly discuss the application of geochemistry in assessing the origin of geothermal fluids, the chemical composition of geothermal fluids from some well-known geothermal fields, and geochemical prospecting.

Origin of geothermal fluids

The origin of geothermal fluids has been debated for a long time. This problem was convincingly solved by Craig et al. (1956) and Craig (1963) through detailed studies of the isotope ratios H/D and $^{16}O/^{18}O$ of geothermal fluids. They determined these ratios from a number of widely separated areas and compared the values with the isotopic composition of meteoric waters from the concerned regions. The H/D ratio in the meteoric and geothermal waters was found to be same for all the sites. However, the $^{16}O/^{18}O$ ratio was found to differ. This is explained as due to exchange of the oxygen isotopes in the thermal water with the oxygen of the wall rocks through which the geothermal water circulated. This proved that geothermal water is of meteoric origin. However, it may contain a small percentage (5—10%) of juvenile water. Recent studies of rare geothermal gases suggest that they are mostly of atmospheric origin. However, excess of ^3He in some systems may come from the mantle (Craig et al., 1975). The mantle contribution of ^3He isotope does not necessarily indicate that other mantle-derived components are also present in the geothermal fluids.

Chemical composition of geothermal fluids

A large number of chemical analyses of thermal fluids have been published so far; they have been summarized by Truesdell (1976). In Table 5-1 some selected data are presented. The geothermal systems type, in this table, is designated following classifications proposed by Ivanov (1967), White (1970), Arnórssón (1974), Kononov and Polak (1976) and Mahon (1976). With the help of these classifications, the volcanic systems where the heat source is a recent magma intrusion in the shallow crust, are distinguished from non-volcanic systems — where the heat source is the normal or elevated geothermal gradient and the fluids are heated up due to deep circulation.

Sigvaldason (1973) has summarized the following most important types of thermal waters.

Sodium chloride water. Sodium chloride water has a pH close to neutral at depth. When this type of water approaches the surface, it loses steam and carbon dioxide and consequently becomes slightly alkaline. Chloride is the most common anion present and the ratio of chloride to sulphate is high. The principal cation present is sodium. This is the most common type of

The following abbreviations are used:

System type: VW = volcanic hot water system; VS = volcanic steam (vapor-dominated system); VSw = volcanic system involving seawater.

Sample type: s = spring; w = well.

Sample temperature is the surface temperature for a spring or a non-boiling well discharge, the temperature of steam separation for well discharges above boiling, or the downhole temperature if a downhole sampler was used or if the analysis was recalculated to downhole conditions.

Analyses: w = water analysis with all major ions and SiO_2; pw = partial water analysis; pg = partial gas analysis; tr = trace water analysis; g = gas analysis; i = water (^{18}O, D) or other isotopes; T = tritium.

Water type is calculated on a weight basis. The symbol means, A = B: A approximately equals B in concentration; A ⩾ B: A is 1 to 1.2 times the concentration of B; A > B: A is 1.2 to 3 times the concentration of B; A >> B: A is 3 to 10 times the concentration of B; A >>> B: A is more than 10 times the concentration of B.

TDS is the sum of the reported constituents of the analysis in ppm (mg/kg).

Gases are in order of molar or volume abundance with the same symbols as for water type.

t_{SiO_2} *adia.*: the quartz saturation temperature (°C) assuming maximum steam loss during cooling (adiabatic cooling) calculated by the computer program GEOTERM (Truesdell, 1976, p. 831), along with t_{SiO_2} cond., $t_{Na\text{-}K\text{-}Ca}$, WSMM, and BSMM which are defined below. No allowance has been made for dissociation of dissolved silica. Some spring systems have data indicated as ave. (average), max. (maximum) and σ standard deviation.

t_{SiO_2} *cond.*: the quartz saturation temperature assuming no steam loss during cooling (conductive cooling).

$t_{Na/K}$: the temperature calculated from the ratio of Na to K using the White-Ellis curve.

$t_{Na\text{-}K\text{-}Ca}$: the Na-K-Ca temperature calculated using Na-K-Ca equation.

Other geothermometers

Na-K-Ca-CO_2: the Na-K-Ca geothermometer with correction applied for high CO_2 contents (Paces, 1975).

chalc: the chalcedony saturation geothermometer with conductive cooling.

$CaSO_4$ sat.: temperature calculated for saturation of anhydrite.

WSMM: the warm spring mixing models described in the text with no steam loss before mixing. Where no other data were available, the cold water component temperature was estimated as equal to the mean annual temperature and the SiO_2 content was assumed to be 25 ppm.

BSMM: the boiling spring mixing model described in the text. The cold spring temperature was estimated as above and the Cl contents estimated (in the absence of data) as 2 to 15 ppm according to the distance from the ocean.

$\Delta^{13}C$ (CO_2, CH_4): temperatures indicated by the fractionation of ^{13}C between CO_2 and CH_4. The notation for this and other isotope geothermometers is self-evident.

$K(CO_2 \rightarrow CH_4)$: temperature calculated from chemical equilibrium constants for the reaction $CO_2 + 4H_2 = CH_4 + 2H_2O$.

Boiling calc.: temperature calculated from the apparent increase in concentration of seawater due to boiling.

Na-Ca-SiO_2, isotope mixing, ^{14}C mixing, "isotope", heat balance, Cl-E: special methods explained in the original references.

Observed temperature is aquifer temperature rather than maximum temperature where aquifers are identified; otherwise, maximum recorded temperature.

References in many cases are grouped where data for a well, spring, or geothermal system are from more than one source.

TABLE 5-2

Chemical composition of typical geothermal waters (modified from Sigvaldason, 1973)

	Sodium chloride water[1]	Acid sulphate/ chloride water[2]	Acid sulphate water[3]	Calcium bi-carbonate water[4]
SiO_2	640.0	412.0	109	60
Li	14.2	n.d.	n.d.	2.3
Na	1320.0	609.0	2.0	129.0
K	225.0	51.0	3.0	69.0
Rb	2.8	n.d.	n.d.	—
Cs	2.5	n.d.	n.d.	—
Ca	17.0	14.0	2.2	272.0
Mg	0.03	4.0	—	68.0
F	8.3	—	—	2.4
Cl	2260.0	878.0	15.0	170.0
Br	6.0	—	—	—
I	0.3	—	—	—
SO_4	36.0	262.0	758	501.0
As	4.8	—	—	—
B	28.8	4.4	6.9	4.3
NH_3	0.15	2.0	30.0	1.0
HCO_3	19.0	—	—	667.0
CO_3		—	—	—
H_2S		—	—	2.6
pH	8.6	3.1	1.97	6.6
T (°C)		55	90	72

[1] Wairakei, New Zealand (Ellis and Mahon, 1964).
[2] Frying Pan Lake, Tarawera, New Zealand (White et al., 1963).
[3] Norris Basin, Yellowstone Park, U.S.A. (White et al., 1963).
[4] Mammoth, Yellowstone Park, U.S.A. (White et al., 1963).
n.d. = not determined.

water present in large underground reservoir systems. The geothermal water from Wairakei, New Zealand (Table 5-2, column 1; Ellis and Mahon, 1964) is a typical example.

Acid sulphate/chloride water. This type of water is relatively uncommon. Oxidation of sulphide to bisulphate at depth causes acidity. The pH shifts from neutral to acid as the water rises to the surface and cools. Frying Pan Lake, Tarawera, New Zealand (Table 5-2, column 2; White et al., 1963) is a typical example of this kind of water.

Acid sulphate water. This kind of water is most common in fumarolic areas where steam rising from hot water reservoirs condenses as it approaches the surface. The H_2S content of the steam, on contact with the air, is converted into H_2SO_4 through oxidation. The chloride content of water is low and it may contain large and varying amounts of cations derived by acid leaching of rocks. The discharge of springs characterized by acid water is usually small and the water often contains suspended clay material. The waters from Norris Basin, Yellowstone Park, U.S.A. (Table 5-2, column 3; White et al., 1963) are a typical example.

Calcium bicarbonate water. This kind of water normally occurs as low-temperature travertine-depositing springs. Calcite precipitation and low temperatures render the use of calcium bicarbonate water difficult. The waters from Mammoth, Yellowstone Park, U.S.A. (Table 5-2, column 4; White et al., 1963) are a typical example.

Volcanic water systems are usually characterized by waters of neutral sodium chloride type at depth, which could be altered during its passage to the surface by addition of calcium, bicarbonate or acid sulphate components. The subsurface temperatures and the concentration of soluble components in the rocks affect the amount of soluble components like Cl, B, Br, Li, Cs, and As observed in the thermal waters. The presence of less-soluble components like SiO_2, Ca, Mg, Rb, K, Na, SO_4, HCO_3 and CO_3 is controlled by temperatures at depth, mineral solubility, pH value and mineral equilibria. The chloride content is usually low in the near-surface fluids of vapor-dominated systems. They normally contain H_2S, CO_2, HBO_2, Hg, NH_3, etc., which are soluble in low-pressure steam. For non-volcanic geothermal systems, the water composition varies widely from dilute meteoric waters to connate waters, oil field brines, metamorphic water, etc. Relatively much less is known about the factors controlling the composition of waters from non-volcanic geothermal systems.

In addition to geothermal waters, analysis of gases escaping from geothermal areas is useful in understanding subsurface conditions. According to Kononov and Polak (1976), fluids related to volcanic processes are characterized either by H_2S-CO_2 gases and acid sulphate or acid sulphate/chloride waters in the oxidizing zone, or by N_2-CO_2 gases and alkaline sodium chloride waters in the reducing zone. Fluids connected with thermometamorphic processes have high concentrations of CO_2 gases and carbonate waters. Geothermal fluids originating from deep circulation, outside volcanic and thermometamorphic zones, have high concentrations of N_2 gases and dilute sodium chloride/sulphate waters.

It is not uncommon to find more than one kind of fluid occurring at the same geothermal system. It is necessary to find a logical relationship between the different kind of fluids and to interpret from the chemistry of the geothermal fluids, the chemical composition of fluids at depth as well as the physical environment of the geothermal reservoir.

Geothermometers

Important parameters like subsurface temperature and flow patterns, source of recharge, and type of reservoir rock can be estimated through chemical and isotopic analysis of fluids reaching the surface of springs or wells. Most important is the relation between hot-spring chemistry and reservoir temperature. Component concentration and ratio of certain elements can be related to subsurface temperatures. These are called geothermometers. Chemical geothermometers can be quantitative — which are used for estimating specific temperatures at depth, or qualitative — which are used to estimate relative temperatures.

Quantitative geothermometers

Silica geothermometers are one of the oldest and most commonly used types. Geochemical studies have shown that quartz is an important secondary mineral phase present and therefore it is common to compare the silica value of the spring water with the quartz solubility curve. One of the assumptions made in such calculations is that silica dissolved at high temperature at depth remains metastable in the solution and silica does not precipitate as the thermal waters rise to the surface. This is mostly true when the spring discharge is considerable and water rises rapidly. On the other hand, higher-temperature values will be inferred if additional silica is dissolved from the wall rocks. The silica geothermometer therefore needs to be used with caution. The subsurface temperatures can be estimated using relations of the type given by Truesdell (1976),

quartz, adiabatic cooling (SiO_2 in ppm):

$$\theta°C = \frac{1533.5}{5.768 - \log SiO_2} - 273.15 \tag{5-1}$$

quartz, conductive cooling (SiO_2 in ppm):

$$\theta°C = \frac{1315}{5.205 - \log SiO_2} - 273.15 \tag{5-2}$$

chalcedony, conductive cooling (SiO_2 in ppm):

$$\theta°C = \frac{1015.1}{4.655 - \log SiO_2} - 273.15 \tag{5-3}$$

In accordance with the first relation, if 200 ppm of SiO_2 are measured in a particular spring water sample, the expected temperature of the reservoir would be 169°C.

Another commonly used geothermometer is the atomic ratio of sodium to potassium. A complicated equilibrium involving alkali-feldspar and potassium-mica controls this ratio. There is a wide divergence in calibration scales and the empirical calibration of this geothermometer does not agree with experimental studies of feldspar and mica equilibrium. Syntheses of available data by White (1970) and Fournier and Truesdell (1973) have produced the following two slightly different scales (adopted from Truesdell, 1976). In these scales Na and K are measured in ppm.

$$\theta°C = \frac{855.6}{\log(Na/K) + 0.8573} - 273.15 \quad (5\text{-}4)$$

$$\theta°C = \frac{777}{\log(Na/K) + 0.70} - 273.15 \quad (5\text{-}5)$$

The first relation is more commonly used. According to it, if the Na/K ratio is found to be 10 in a certain spring water, the subsurface temperature is expected to be about 202°C. The Na/K geothermometer fails at temperatures lower than 100 to 120°C and gives high temperatures for solutions with high calcium contents. The Na/K ratio has longer metastable persistence than SiO_2. However, potassium is preferentially absorbed by certain hydrothermal clay minerals, such as montmorillonite, present in the top few hundred meters of the layers in geothermal areas. Therefore, the Na/K thermometer should be used with caution.

Another which has been recently used is the Na-K-Ca geothermometer (Fournier and Truesdell, 1973, 1974). These authors found the relation:

$$\theta°C = \frac{1647}{\log(Na/K) + \beta \log(\sqrt{Ca}/Na) + 2.24} - 273.15 \quad (5\text{-}6)$$

where Na, K and Ca are measured in moles/liter and:

$$\beta = 4/3 \text{ for } \sqrt{Ca/Na} > 1 \qquad \theta < 100°C \quad (5\text{-}7)$$

$$\beta = 1/3 \text{ for } \sqrt{Ca/Na} < 1 \qquad \theta > 100°C \quad (5\text{-}8)$$

Both of the cation geothermometers, i.e. Na/K and Na-K-Ca, depend on equilibria between thermal waters and the alumino-silicate minerals of the host rock. Therefore, misleading temperatures are inferred if equilibrium is not reached or some unusual suite of minerals exists. As pointed out by Truesdell (1976), cation geothermometers should be used with caution in geothermal systems involving seawater, because in such systems equilibrium is often not reached due to resistance to chemical changes.

Absence of magnesium in geothermal waters is an indication of high reservoir temperatures. Hydrothermal experiments show that magnesium is preferentially incorporated in clay minerals. Clay minerals, such as chlorite, are stable at high temperatures. Therefore, geothermal water at high temperatures, say above 200°C, in contact with altered rocks is depleted in magnesium (less than 1 ppm).

Mixing models. Very often the cold water from near the surface gets mixed with the thermal waters. This makes the direct application of the chemical geothermometers difficult. Under favorable conditions, the original temperature of the hot water and the fraction of the cold water in the mixture can be estimated by measuring the temperature and silica content of the warm

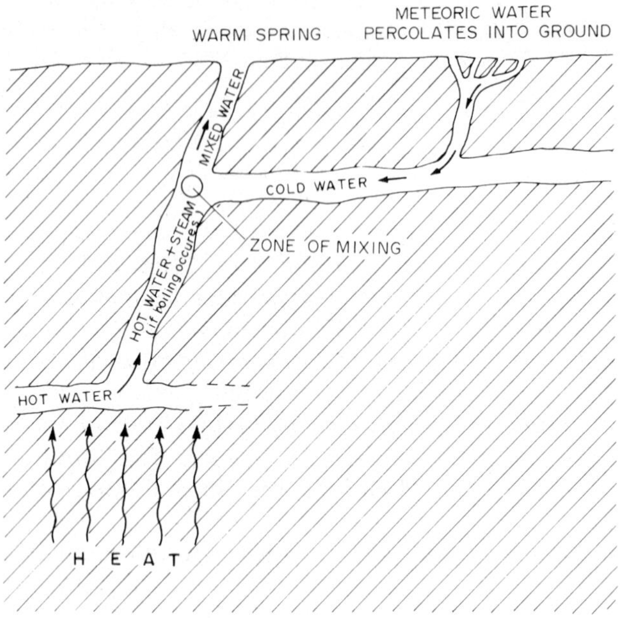

Fig. 5-2. A mixing model (from Fournier and Truesdell, 1974).

spring water, as well as the temperature and silica content of non-thermal water in the region. Fig. 5-2, from Fournier and Truesdell (1974), shows a typical situation. Hot water ascends from depth along a permeable channel, and at some point it encounters cold water. At the depth of mixing, the weight of the column of the cold water extending to the surface is greater than the weight of the hot water column to the surface. Cold water therefore enters the hot water channel and the mixture flows to the surface and is discharged as a warm spring. To determine the two unknowns, i.e., the temperature of the hot water and the proportion of the hot and cold waters, two equations can be written and solved. The first equation relates the enthalpies of the hot water (H_h), cold water (H_c) and spring water (H_s) and the fractions of cold water, X, and of hot water $(1 - X)$ as follows:

$$(H_c)\,X + H_h(1-X) = H_s \qquad (5\text{-}9)$$

The second equation relates the silica contents of hot water (Si_h), cold water (Si_c), and spring water (Si_s):

$$(Si_c)\,X + Si_h(1-X) = Si_s \qquad (5\text{-}10)$$

These two equations can be solved to obtain the two unknowns. Fournier and Truesdell (1974) have suggested a graphical method for determining the unknown as follows:

(1) A series of values of the enthalpy of hot water for temperatures listed in Table 5-3 are assumed and X_t is calculated for each value as follows:

TABLE 5-3

Enthalpies of liquid water and quartz solubilities at selected temperatures and pressures (Fournier and Truesdell, 1974)

Temperature (°C)	Enthalpy (cal/g)	Silica (mg/l)
50	50.0	13.5
75	75.0	26.6
100	100.1	48.0
125	125.4	80.0
150	151.0	125.0
175	177.0	185.0
200	203.6	265.0
225	230.9	365.0
250	259.2	486.0
275	289.0	614.0
300	321.0	692.0

$$X_t = \frac{(\text{enthalpy of hot water}) - (\text{temperature of warm spring})}{(\text{enthalpy of hot water}) - (\text{temperature of cold spring})} \qquad (5\text{-}11)$$

(2) X_t is plotted against the temperatures from which the assumed hot water enthalpy values are derived (curve *A* in Fig. 5-3).

(3) A series of values of the silica contents of hot water for the temperatures listed in Fig. 5-3 is assumed and X_{Si} is calculated for each value as follows:

$$X_{Si} = \frac{(\text{silica in hot water}) - (\text{silica in warm spring})}{(\text{silica in hot water}) - (\text{silica in cold spring})} \qquad (5\text{-}12)$$

(4) X_{Si} is plotted against temperature on the same graph (curve *B* in Fig. 5-3). The point of intersection of these two curves gives the estimated temperature of the hot-water component and the fraction of the cold water. In Fig. 5-3, curves *A* and *B* are drawn for a case where the warm spring water and cold water temperatures are 60°C and 10°C, respectively, and the corresponding silica contents are 45 and 5 mg/l, respectively. The intersection of *A* and *B* shows the cold water fraction to be 0.64 and the temperature of the hot water is inferred to be 148°C.

At times these two curves do not intersect, or intersect at an unreasonably high temperature. These situations arise when the ascending hot water loses heat before mixing with cold water and a different procedure is suggested by Fournier and Truesdell (1974) to deal with such a situation. The accuracy of the mixing model depends to a large extent on the exactness with which the chemistry and temperature of the local cold subsurface waters are determined. Recently Gupta et al. (1976a) and Young and Whitehead (1975a, b) have successfully applied warm spring mixing models for estimating subsurface reservoir temperatures.

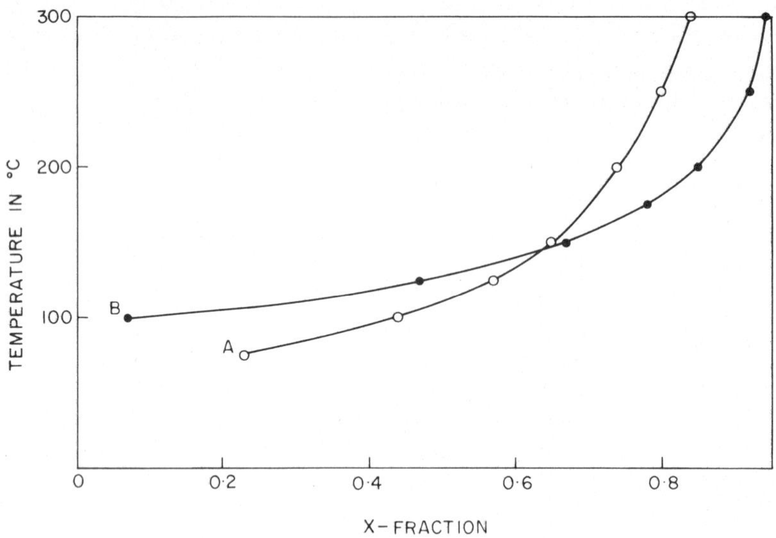

Fig. 5-3. Relations between fraction of cold water and temperature in a mixing model. Fraction of cold water based on model shown in Fig. 5-2 with enthalpy consideration (*A*), and with silica content consideration (*B*). For details see text.

Qualitative geothermometers

As the name implies, qualitative geothermometers are indicators of subsurface geothermal regimes, but they cannot be used to infer subsurface temperatures. They are most useful in preliminary exploration over large areas. In many areas surface fluid flow may not be conspicuous enough to indicate a subsurface geothermal potential. However, chemical analyses of surface samples may reveal the presence of certain distinctive geothermal components. We shall refer here to a few surveys and analyses reported in the literature where some of the indications have been discussed.

Koga and Noda (1976) report that sensitive analytical methods have shown that soil gases in geothermal areas have high concentrations of mercury. Over the Beppu geothermal system in Japan, the atmosphere is found to contain an anomalously high concentration of mercury vapor (Koga and Noda, 1976). The presence of helium (Roberts et al., 1975) and carbon dioxide with anomalously high $^{13}C/^{12}C$ ratios (Rightmire and Truesdell, 1974) have been also reported from geothermal areas. Brondi et al. (1973) have found the presence of lithium in surface waters in central Italy to be an indication of subsurface geothermal potential. Dall'Agilo and Tonani (1973) have developed a criterion for distinguishing river sulphate of geothermal origin, produced by H_2S oxidation, from sulphates resulting from solution of evaporites or from the oxidation of sulphide minerals. The presence of anomalous boron in surface waters is also considered to be a geothermal indicator (Morgan, 1976). Steam emerging from the Larderello geothermal fields in Italy exhibits a high boron content. Natural and exploited geothermal systems in the Waikalo River, New Zealand, show high percentages of arsenic (Rothbaum and Anderton, 1976). Anomalous arsenic has been also reported from the Madison River, Montana, U.S.A. (Stauffer and Jenne, 1975). Fancelli et al. (1975) and Fouillac et al. (1976) have used the *oxygen shift* as a positive or negative qualitative geothermometer. This oxygen shift is caused when geothermal waters of meteoric origin exchange oxygen isotopes with rocks during deep circulation.

Isotope geothermometers

Isotope geothermometers, in general, equilibrate more slowly than chemical geothermometers and are capable of indicating the temperatures of the deeper parts of geothermal systems. Much new information was presented and discussed in an advisory group meeting on the application of nuclear techniques to geothermal studies (Gonfiantini and Tongiorgi, 1976). Truesdell (1976) has summarized the uncertainties associated with various isotope geothermometers.

The distribution of carbon isotopes, ^{13}C, between carbon dioxide and methane [$\Delta^{13}C$ (CO_2, CH_4)] has been used as geothermometer at Larderello, Italy (Ferrara et al., 1963); Wairakei and Broadlands, New Zealand (Hulston and McCabe, 1962; Lyon, 1974) and elsewhere. At Larderello and Wairakei, the temperatures estimated using the carbon isotope technique were in good agreement with the actually measured reservoir temperatures. However, those inferred at Broadlands were considerably higher than the actual reservoir temperatures. More experimental work needs to be carried out on this geothermometer to assess various factors affecting temperature estimates. A hydrogen isotope geothermometer [$\Delta D(H_2, CH_4)$] has been recently calibrated (Horibe and Craig, in Craig, 1976). Both the hydrogen isotope geothermometers, i.e., $\Delta D(H_2, CH_4)$ and $\Delta D(H_2, H_2O)$, have been tested earlier at a few places (for example at Imperial Valley, California (White, 1968); Olkaria, Kenya (Glover, 1972, 1973)), they appear to re-equilibrate quickly in most cases.

The other isotope geothermometers which need to be mentioned are the $\Delta^{34}S(SO_4, H_2S)$ and $\Delta^{13}C(CO_2, HCO_3)$ geothermometers. The former was calibrated by Robinson (1973); however, it yielded much higher temperatures at Wairakei, New Zealand (Kusakabe, 1974). The latter has been tested at Yellowstone Park, Wyoming (Truesdell and Fournier, 1976) and elsewhere. Both these geothermometers, however, need more testing.

GEOPHYSICAL TECHNIQUES

Geophysical methods in geothermal exploration have been reviewed periodically (i.e., Banwell, 1970, 1973; Bodvarsson, 1970; Combs and Muffler, 1973; Combs, 1976; Palmason, 1976). Some of the geophysical methods are now quite well established, having been successfully used in geothermal exploration projects. Other new methods are being developed and tested. Exploration work in geothermal areas is mostly aimed at delineating the geometry of shallow geothermal reservoirs with upflow zones causing geothermal manifestations. However, deeper systems exist without any up-flow zones and surface manifestations. The most suitable geophysical methods for exploring shallow reservoirs differ in part from those used for deep reservoirs. The thermal state of the fluid in the reservoir adds further complications. The physical properties of porous rocks, such as resistivity and density, are different in rocks filled with steam from those filled with hot water.

Various classifications of the geophysical methods used in geothermal exploration have been adopted (Bodvarsson, 1970). The methods can be classified according to the depth range for which they are particularly suited. They can also be classified as direct and indirect methods (Palmason, 1976). Direct methods include thermal exploration techniques which aim at mapping thermal zones geometrically. Indirect methods include techniques for investigating geological structures that may control the movement of geo-thermal fluids. In general, these classifications have considerable overlap. Following Palmason (1976), thermal and electrical methods are discussed first and then the structural methods. Other techniques, which do not fall into these two major categories are discussed later.

Thermal methods

Thermal exploration techniques are extremely useful in assessing the size and potential of a geothermal system. Surface temperature, temperature gradient and conductive heat flow measurements are often made in geothermal energy exploration and used as a primary criterion for selection of a drilling site. The average global conductive heat flow is 1.5 HFU. In combination, or individually, factors such as high content of radioactive material, exothermic chemical reactions, movement along faults and other similar factors can cause heat flows slightly in excess of the normal values. Geothermal areas which are economically attractive at the present time have heat flows several times higher than the global average.

Thermal methods of prospecting normally make use of the following three measurements.

One-meter temperature probe surveys

One-meter depth-temperature measurement surveys are inexpensive and quick, and can be used to detect anomalously hot areas. When geological structure controls thermal activity, one-meter probe surveys have the same value as infra-red imagery surveys. Measurements are made along profiles or in a grid pattern at distances varying from 10 to 100 m. A simple probe (Fig. 5-4C) is used. Due care must be taken of near-surface effects such as topography, precipitation, insolation, and movement of groundwater, in interpreting the results of 1-m depth probe surveys. The effect of groundwater movement is most important. Relatively slowly moving groundwater can carry away conductive heat, even over a strong geothermal anomaly, and grossly distort the pattern of surface temperature. McNitt (1976) reports that 1-m temperature probe surveys have been undertaken in almost all U.N.-sponsored geothermal projects and have discovered very little that was not already known, with the exception of the ones carried out in areas of steaming ground. The most compelling reasons for undertaking these surveys are their being inexpensive and simple and the fact that they may discover something new.

Geothermal gradient measurements

Temperature gradient measurements have been often used as a primary criterion in selecting drilling sites. A thermal probe used for such measurements is shown schematically in Fig. 5-4A. Most commonly, temperature gradient measurements are made in boreholes at depths varying from 10 to 150 m. At these depths, the surficial thermal disturbances are quite negligible and measurements can be made quite reliably. However, it is necessary to correct the data for lateral groundwater movements. The average geothermal gradient at shallow depths is about 3°C/100 m, whereas in potential geothermal areas it is found to be more than 7°C/100 m (Combs and Muffler, 1973). Geothermal gradient measurements are useful in defining the areal extent of thermal anomalies. However, care should be exercised in extrapolating the geothermal gradients measured at shallow depths to deeper regions. The linear extrapolation of thermal gradients is found to be erroneous, giving high values, due to decrease in porosity with depth and the effect of convection cells at depth. Particularly in sedimentary rocks, porosity decreases considerably with depth. The thermal conductivity of minerals is

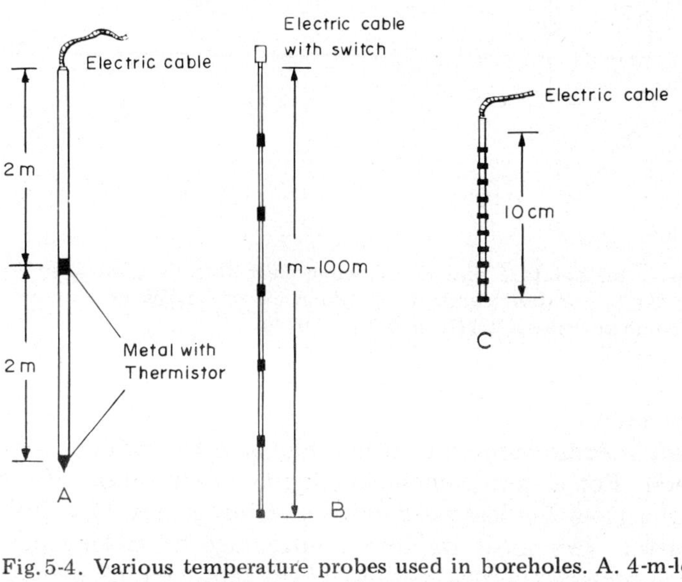

Fig. 5-4. Various temperature probes used in boreholes. A. 4-m-long rigid probe with two thermometers. B. Cable probe with a number of thermometers. C. Short probe for near-surface measurements.

much higher (3 to 10 times) than that of water. Consequently, with the decrease of porosity, bulk thermal conductivity increases considerably and the gradient decreases. The convection cells have an even larger contribution in decreasing temperature gradients at depth. Fig. 5-5, from White (1973), shows that convection produces higher thermal gradients at the top of the convection cell and low thermal gradients in the region within the convection cell.

Albright (1976) has reported a relatively new and accurate method of thermal gradient measurement in deep bore holes. In this method, equilibrium rock temperatures are calculated with high precision from an analysis of the thermal recovery of the bottom hole temperature measured continuously. A thermistor sonde (Dennis and Todd, 1975) is used for making bottom hole temperature measurements. This information is used in determining thermal gradients. Thermal gradients were measured using this technique for the first time on the western flank of Valles Caldera in north-central New Mexico (Albright, 1976).

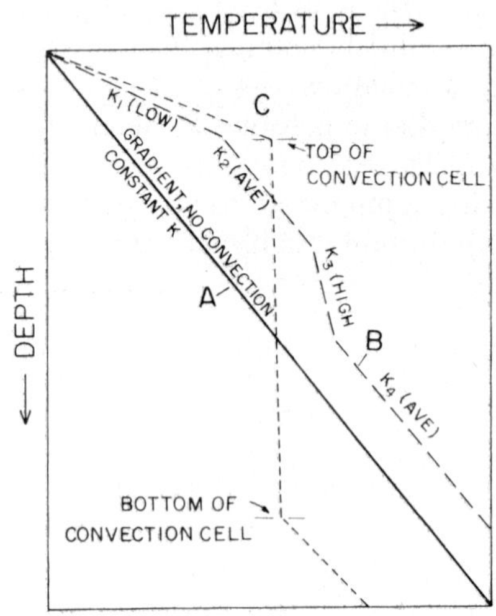

Fig. 5-5. Relations between temperature and depth when heat flow is controlled by thermal conduction in rocks of constant conductivity (*A*), rocks of variable conductivity (*B*), or by major convective disturbances (*C*) (from White, 1974).

Heat flow measurements

Geothermal gradient measurements are often adequate to outline a geothermal area in general. For a better understanding and delineation of the central production zone from the less productive marginal zones, heat flow measurements are useful. The most important advantage of making heat flow measurements compared to purely geothermal gradient measurements is that heat flow measurements are independent of the in-situ thermal conductivity of the rocks. Therefore, as explained by Sestini (1970), in inhomogeneous terrain heat flow measurements alone can provide accurate information regarding productive zones. One of the earliest demonstrations of the usefulness of heat flow measurements in delineating the central production zone was given by Boldizsár (1963) in the natural steam field at Larderello, Italy.

The distance between boreholes used for heat flow measurements depends upon the size of the subsurface heat source. Magmatic intrusions, which are usually the sources for economic geothermal fields, cause geothermal disturbances of at least 1—2 km lateral extent. Therefore, measurements at distances of 500 m are considered to be adequate. Since heat flow is a product of thermal gradient and thermal conductivity, it is necessary to measure the thermal conductivity of the formations using cores obtained by drilling. A typical apparatus used for measurement of the thermal conductivity is shown in Fig. 5-6 (after Creutzburg, 1964). The sample, usually in a cylindrical form, is placed next to the heater H and necessary precautions are taken that heat should be completely conducted through the sample. No heat should be lost and no heat should be permitted to be transmitted to the

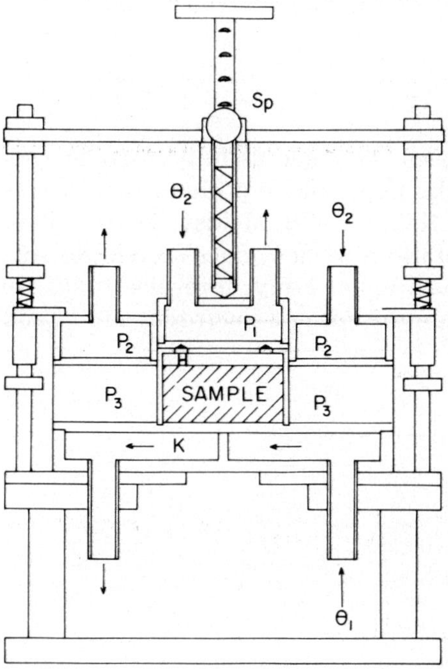

Fig. 5-6. Schematic representation of a thermal conductivity measurement apparatus.

sample. Protectors *P1*, *P2* and *P3* are provided to achieve this. When the steady state is reached, conductivity K is given by:

$$K = \frac{0.239\, I^2 R h}{F(\theta_2 - \theta_1)} \qquad (5\text{-}13)$$

where I = current in the heater H (A), R = resistance in the heater (Ω), h = thickness of the sample (cm), F = area of the sample (cm^2), θ_2 = temperature at the bottom of the sample, and θ_1 = temperature at the top of the sample.

Heat flow measurements are considered to be reliable when boreholes are at least 100 m deep. Many other necessary precautions and procedural details have been given by Sestini (1970), Burgassi et al. (1970) and others. In recent years, geothermal methods have been successfully used by Cermak et al. (1976), Franko and Mucha (1976), Boldizsár and Korim (1976), Gupta et al. (1976b) and others, as an indicator of hydrothermal activity at depth and in assessment of the potential of geothermal regions in various parts of the world.

In a novel study conducted many years ago, White (1969) estimated heat flow in Yellowstone National Park, U.S.A. utilizing individual snow falls as calorimeters. After a favorable snow fall — characterized by being of brief duration, heavy, and occurring when the air temperature was close to 0°C with no wind — contacts between snow-free and snow-covered ground were mapped at a suitable scale. Each mapped contact, as time passed, represented a heat flow contour.

Electrical resistivity methods

Electrical resistivity methods have been very successfully used in geothermal exploration as is evident from the numerous reported case histories in the literature (e.g., Risk et al., 1970; Keller, 1970; Meidav, 1970; Zohdy et al., 1973). The three classical configurations of electrodes used in resistivity surveys are shown in Fig. 5-7. In the Wenner array, four electrodes are equally spaced at intervals a. C_1 and C_2 are the current electrodes and P_1 and

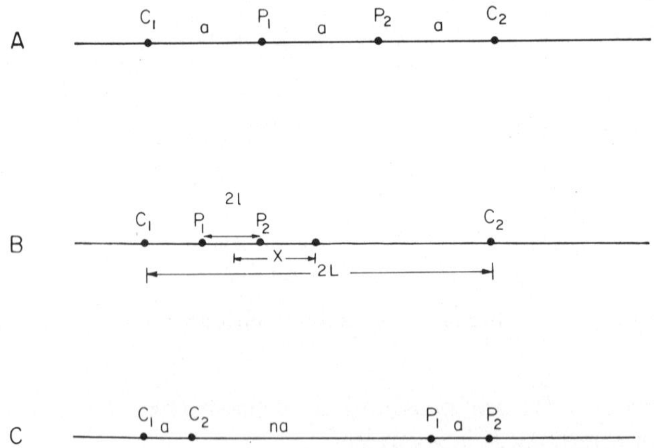

Fig. 5-7. Some commonly used electrode configurations in electrical resistivity surveys: (A) Wenner, (B) Schulmberger, and (C) dipole-dipole configurations.

P_2 are potential electrodes. For this configuration, apparent resistivity ρ_a is given by:

$$\rho_a = 2\pi a \frac{\Delta V}{I} \qquad (5\text{-}14)$$

where ΔV is the potential difference between P_1 and P_2 when current I is sent through the current electrodes. The distance between potential electrodes, $2l$, in the generalized Schlumberger array is small compared to the distance, $2L$, between the current electrodes. When $L \geqslant 5l$, the apparent resistivity can be estimated using the relation:

$$\rho_a \simeq \frac{\pi}{2l} \left[\frac{(L^2 - X^2)^2}{L^2 + X^2} \right] \frac{\Delta V}{I} \qquad (5\text{-}15)$$

where X is the distance between the center of the current electrodes and the center of the potential electrodes. In the symmetrical Schlumberger array arrangement, $X = 0$, and apparent resistivity is estimated using the relation:

$$\rho_a \simeq \frac{\pi L^2}{2l} \frac{\Delta V}{I} \qquad (5\text{-}16)$$

In the dipole-dipole configuration of electrodes, the potential electrodes are placed outside the current electrodes, each pair having a constant mutual separation, a. When the distance between the two pairs, na, is large (Fig. 5-7), the current source can be treated as a dipole and the apparent resistivity is estimated using the relation:

$$\rho_a \simeq \pi n(n+1)(n+2)\, a \frac{\Delta V}{I} \qquad (5\text{-}17)$$

The above-mentioned configurations are the standard ones. Various others have been tested in recent years to increase the depth of penetration when the near-surface resistivities are low — a situation mostly encountered in hydrothermal areas.

Resistivity *sounding* and *profiling* are two commonly employed procedures for estimating underground resistivity. The object of resistivity sounding is to estimate variation of resistivity with depth below a given point. Such measurements are required when the ground consists of a number of more or less horizontal layers and vertical variation in resistivity is to be estimated. With increasing separation of the current electrodes, the current penetrates deeper. While using the Wenner configuration for resistivity sounding, the electrode separation a is increased in steps, for example $a = 4$, 8, 16, 32 m. The object of resistivity profiling is to detect the lateral variation of resistivity of the ground. This kind of survey is undertaken to delineate underground intrusions. In electrical profiling with the Schlumberger method, the current electrodes are fixed at a large separation and the potential electrodes are moved at a constant small separation within the current electrodes (Fig. 5-7). Profiling is normally conducted at right angles to the strata of the structure such as faults and dykes. Standard master curves are available in the literature (for example, those published by the European Association of Exploration Geophysicists, 1963) to interpret field observations.

Various factors contribute in increasing the electrical resistivity contrast between the geothermal systems and the surrounding rocks. Firstly, the electrical resistivity decreases with the increase of temperature. Secondly, with the increase of temperature, solubility increases and consequently there is an increase in salinity and decrease in resistivity. Thirdly, the pore space in the central portion of a geothermal field increases with increase in temperature, thereby further decreasing the resistivity. All these factors contribute to decreasing the resistivity of geothermal fields relative to the surrounding rocks; in the classical geothermal systems, the electrical resistivity is reported to be 5 Ω-m, while that of the surrounding rocks is in excess of 100 Ω-m (Meidav, 1972).

As summarised by Meidav and Tonani (1976), there are three major factors which introduce ambiguity and electrical resistivity methods cannot always be used as a perfect geothermal exploration tool. These are (1) the effect of dry steam, (2) the effect of brine, and (3) the presence of highly porous but non-permeable rocks like clay and shale.

In liquid-dominated geothermal reservoir systems, the resistivity decreases with the increase of temperature. This is not the case with dry-steam-dominated geothermal reservoirs. Investigations carried out at vapor-dominated geothermal systems at Larderello, Italy (Battini and Menut, 1964), The Geysers, California (Stanley et al., 1973) and Matsukawa, Japan (Hayakawa, 1966) have failed to show low resistivity. When a non-condensable gas cap overlies a boiling water table, resistivity soundings have revealed a high-resistivity layer sandwiched between two layers of low resistivity. Fig. 5-8, adapted from Meidav and Tonani (1976) shows a model of the type of resistivity depth sounding observed in an area where a dry steam layer exists, compared to a region where temperatures increase moderately with depth and the sandwiched vapor-dominated layer is missing.

Cold brine or seawater has an electrical resistivity of less than 1 Ω-m. Therefore many basins, where brine is accumulated, exhibit low resistivity without any connection with elevated subsurface temperatures. Meidav and Furgerson (1972) have reported a classical example from the Imperial Valley, U.S.A., where the resistivity value decreases regionally from 30 Ω-m at Colorado River near Yuma, Arizona, to 1 Ω-m near the Salton Sea, California. These observations indicate an increase in groundwater salinity, in general, in a northwestward direction.

Large sedimentary basins which are rich in clay sand sequences, such as the Gulf Coast of Mexico, are generally characterized by low resistivities. The decrease in resistivity has been caused by the increase in clay or shale content. Therefore, in sedimentary basins, unless there is some other direct evidence, decrease in resistivity cannot be interpreted in terms of the existence of a geothermal reservoir. Meidav and Tonani (1976) have presented a schematic plot of resistivity versus temperature (Fig. 5-9) which is useful in diagnosing the subsurface rocks in terms of their geothermal energy

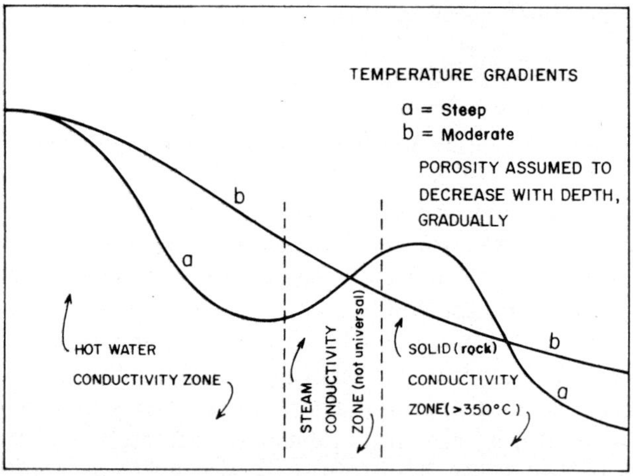

Fig. 5-8. A schematic representation of resistivity-depth graphs for areas where temperature increases moderately with depth (*b*), and for areas where temperature increases rapidly with depth (*a*) where a gas or dry steam layer may exist (from Meidav and Tonani, 1976).

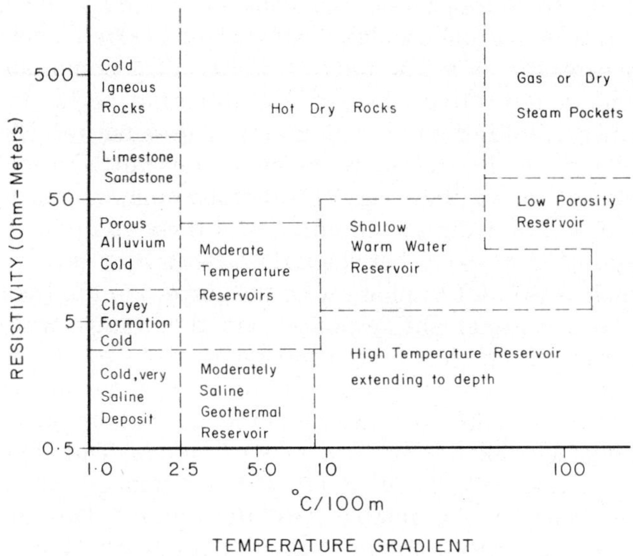

Fig. 5-9. A schematic plot showing the relationship between resistivity and temperature gradients. Such plots are useful in roughly estimating the nature of the subsurface rocks in terms of their geothermal energy potential.

potential. Only a few combinations of high temperature gradient and resistivity are indicative of favorable geothermal conditions. As pointed out by Meidav and Tonani (1976), ambiguity in interpretation of resistivity data can also be minimized by relating the resistivity data with the lithology and salinity of the groundwater. Knowledge of local geological and hydrological conditions is useful in interpreting resistivity data.

Electromagnetic methods

In recent years, electromagnetic methods have been increasingly used in geothermal prospecting (Keller, 1970; Keller and Rapolla, 1974). In principle, the electromagnetic inductive method involves generating a magnetic field which varies in time and detection and measurement of either the electric field or the magnetic field arising from the current induced in the earth. Theoretically, electromagnetic methods have two clear-cut advantages over electrical resistivity methods. Firstly, in the electromagnetic inductive methods signal size increases with decreasing resistivity, making them more suitable for geothermal areas. Secondly, unlike electrical resistivity methods, electromagnetic inductive methods are not adversely affected by the presence of a high-resistivity top layer. The following are the four specific systems/techniques found most useful in prospecting for geothermal resources (Keller, 1970).

Two-loop profiling method

The system consists of two small loops used as a transmitter and a receiver. The transmitter is powered from a small oscillator with selectable frequencies and a power generation capacity of a few tens of watts. Most commonly used frequencies are from a few hundred to a few thousand hertz. The receiver consists of a tuned amplifier and a ratiometer. The signal received through the ground is compared with the reference signal transmitted directly. The relative amplitudes of the in-phase and out-of-phase components are measured as a percentage deviation from the amplitudes of these components when no conductive ground is present between the transmitter and the receiver. The voltage which is 90° out of phase with the current in the transmitter is considered to be a cent-percent response, and the voltage which 180° out of phase with the current in the transmitter is taken to be zero percent response. Typically, the separation between the transmitter and the receiver ranges from 20 to 50 m. The transmitter and the receiver are moved along a profile and readings are taken every few tens of meters. The depth to which conductivity is estimated is about one half the distance between the transmitter and the receiver. The relative readings from a two-loop profiling system can be converted into apparent ground conductivity using some simple relations. When both coils are horizontal, the voltage at the receiver coil is given by:

$$V_c = \frac{\mu \omega A_r M}{2\pi r^2 R^5} \{9 - [9 + 9\nu R + (2\nu R)^2 + (\nu R)^3] e^{-\nu R}\} \qquad (5\text{-}18)$$

where V_c = the complex voltage detected at the receiver coil, ω = the frequency in rad/s, R = the separation between two coils in meters, A_r = the effective area of the receiver coil, M = the moment of the source coil, the product of current and effective area, μ = the magnetic permeability of the earth, usually taken to be $4\pi \times 10^{-7}$ in MKS units, and ν = the complex wave number for uniform earth, usually taken to be $(i\omega\mu\sigma)^{1/2}$, σ being the conductivity in mho/m.

The above-mentioned equation cannot be solved explicitly for conductivity. A curve is drawn from the in-phase and out-of-phase components of V_c, as a function of νR, such as the one shown in Fig. 5-10, and the value of ground conductivity is read from it. Keller (1970) has commented on various features of such a curve. Usually two values of ground conductivity are given from the out-of-phase component. A unique value of ground conductivity is provided from the in-phase component, provided the in-phase response is less than the 100% free air response. When the in-phase response is more than 100%, two values are given. When the response is close to 100%, the value of conductivity can not be determined and a maximum possible conductivity is inferred. When the ground under investigation is uniform, one of

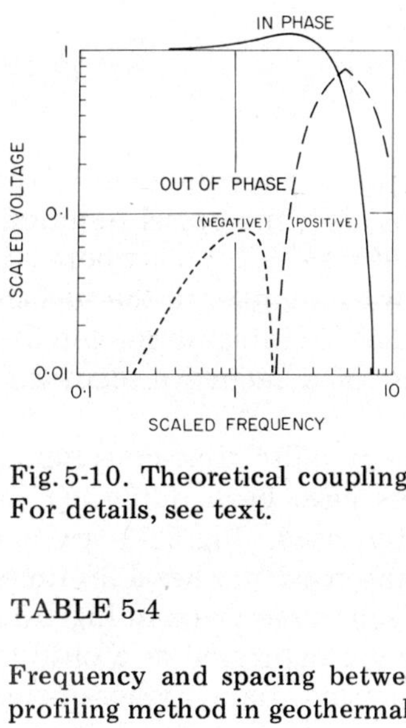

Fig. 5-10. Theoretical coupling curve for a two-loop profiling system after Keller (1970). For details, see text.

TABLE 5-4

Frequency and spacing between the coils suggested by Keller (1970) for the two-loop profiling method in geothermal areas

Frequency (rad/s)	Spacing (m)
10	3800
100	1200
1000	380
10,000	120
100,000	38

the out-of-phase values will agree with the in-phase values. Deviation from this implies that the earth in the vicinity of one of the coils is not uniform. The spacing between the coils and the frequency used should be so chosen that the in-phase response drops to 95% for the lowest conductivity to be resolved. Keller (1970) has suggested a number of combinations of frequencies and spacings for geothermal areas. These are given in Table 5-4.

Two-loop sounding method

In the two-loop sounding method, measurements are made at many frequencies and the data thus generated are interpreted in terms of a number of layers with differing resistivities lying one over the other. With the two-loop profiling method, it becomes difficult to carry the bulky generator used for powering the source, which is necessary to generate the required power at lower frequencies for deeper penetration. This difficulty is somewhat resolved in the two-loop sounding procedure.

Keller and Frischknecht (1966), Vanyan (1967) and others have discussed the theory of the two-loop electromagnetic sounding method. When the earth consists of a number of horizontal layers, the voltage at the receiver can be expressed by a relation of the type (Keller, 1970):

$$V_c = \frac{A_r M \mu}{2\pi R} \frac{\partial}{\partial R} \left[R \frac{\partial}{\partial R} \int_0^\infty \frac{m}{n_0 + n_i/Q} J_0(mR) dm \right] \tag{5-19}$$

where A_r = the effective area of the receiving loop, M = the moment of the source, R = the separation between the two loops, J_0 = the Bessel function, n_0 = the modified wave number in air, given by $n_0^2 = m^2 + \lambda\omega^2\mu\epsilon$ where ϵ is the dielectric constant of air, n_i = the modified wave number in the surface layer given by $n_i^2 = m^2 + i\omega\mu\sigma_i$, m = a dummy variable of integration, and Q = the function, which can be written in the form of a recursive algorithm (Keller, 1970).

Evaluation of the above-mentioned expression is quite time-consuming. However, tables of results for 1, 2 and 3 layers have been published by Frischknecht (1967) which can be conveniently used. Fig. 5-11 gives a typical set of curves for a two-layer case where the top layer has a 30-times higher resistivity than the underlying medium. Each curve given in Fig. 5-11 corresponds to a different thickness of the top layer expressed as a ratio to the distance between the two loops. For interpretation, field measurements are plotted on bi-logarithmic graph paper at the same scale as the reference curve and compared to select a reference curve agreeing closely with the observations. The desired value of conductivity is inferred from the relative positions of the origins on the two plots, and the depth is obtained from the parameters associated with the selected reference curve.

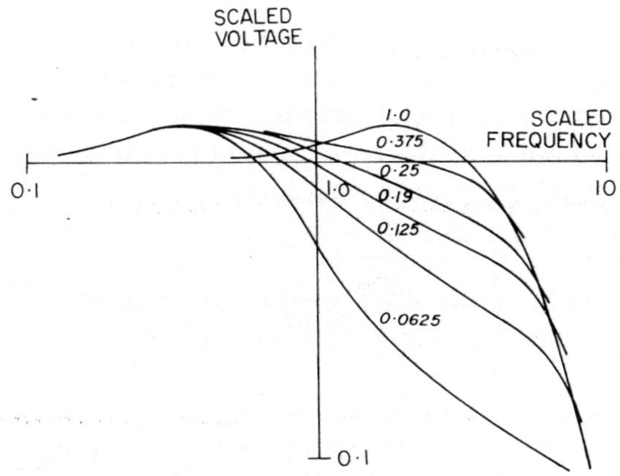

Fig. 5-11. A set of reference curves for electromagnetic coupling between two loops (from Frischknecht, 1967). The curves are valid for a case when a layer rests on a half-space which is 30 times more conductive. For details, see text.

Wire/loop surveys

In these surveys, measurements are made with a short wire source instead of a loop. Moreover, as very low frequencies are to be used for deeper penetration, time domain techniques are preferred to single-frequency methods. This is not a very popular method in geothermal prospecting and hence it is not discussed in detail here. Keller (1970) has given computational and procedural details of this method as used in geothermal prospecting.

Audio-frequency magnetotelluric method

The earlier-described electromagnetic methods require a somewhat cumbersome and bulky source to establish the required moment before measurements can be made. The magnetotelluric method makes use of natural electromagnetic fields (Cagniard, 1953). The audio-frequency magnetotelluric (AMT) method makes use of natural signals in the frequency range of 8 Hz to 20 kHz and it has been found useful in geothermal reconnaissance surveys.

Strangway et al. (1973) have described the AMT method as applied to mineral exploration. AMT exploration makes use of the natural energy originating from world-wide lightning storms. Tropical storm cells, most frequent during the summer months, are the principal contributors to the natural energy. The scalar apparent resistivity is given by:

$$\rho_a = \frac{1}{5f} \frac{|E_x|^2}{|H_y|^2} \tag{5-20}$$

where f = the frequency in hertz, E_x = the horizontal electrical field component in the x direction in 10^{-6} V/m, and H_y = the horizontal component of magnetic field in the y direction in gammas. Apparent resistivity at a given frequency is determined by measuring the mutually orthogonal electric and magnetic fields of the distant lightning. The depth of measurement, δ in meters, is estimated from the relation (Hoover and Long, 1976):

$$\delta = 503\sqrt{\rho/f} \tag{5-21}$$

The instrumental and procedural details of the AMT method in geothermal exploration have been given by Hoover and Long (1976), and Whiteford (1976).

Whiteford (1976) has compared the apparent resistivity measured by the AMT method with those determined by several electrical resistivity surveys carried out in the Broadlands geothermal field in New Zealand. The resistivities obtained by the AMT method were found to be higher. However, the sharp boundaries of the Broadlands geothermal field could be delineated and matched with those deciphered from electrical resistivity surveys.

In general, electromagnetic induction methods are found to be useful in measuring resistivity in hydrothermal areas. The AMT method provides a good and easy to use reconnaissance tool. Controlled-source methods, using a coil or loop, are useful in estimating resistivity at depths ranging from a few tens of meters to a few kilometers. One of the most important applications of induction methods is in areas characterized by surface layers of high resistivity, where direct electrical resistivity measurements are difficult for the deeper layers. The major handicaps of induction methods are the cumbersome equipment and not very direct procedures adopted in determining the resistivity.

Self-potential method

Self-potential anomalies have been reported from a number of geothermal areas. Corwin and Hoover (1979) have reviewed and assessed the application of the self-potential method in geothermal exploration. An analysis of laboratory data and field measurements indicate that self-potential anomalies, comparable to those observed in geothermal areas, can be generated by thermoelectric or electrokinetic coupling processes. When telluric currents and changes in electrode polarization are measured and corrected for, and the self-potential measuring electrodes are not watered, self-potential measurements are reproducible to within ± 5 mV. In arid areas, variations in soil properties can generate reproducible short-wavelength geological noise of as much as ± 10 mV. The amplitude of such noise is much smaller in areas of uniform, moist soil. Since self-potential variations can be caused by changes in geochemical or geological conditions, stray currents from cultural activity, conductive mineral deposits, etc., it is necessary to scrutinize self-potential data carefully before assigning a geothermal origin to them.

In self-potential field surveys, the electrode contact resistance should be checked at each station and efforts should be made to ensure good ground contact. It is necessary to keep the circuit resistance as uniform as possible from one station to another. Measurement of resistance requires driving a current through the electrodes. Therefore, this should be done after taking the self-potential reading and the current should be passed for the shortest possible duration. One of the common electrode configurations used is called the "leap frog" configuration, in which a dipole of fixed length is stepped along the survey line. Successive voltages are added to obtain the total field. Sometimes the addition of small errors accumulates into large values. It is also necessary to monitor the potential generated by telluric currents and to carry out frequent checks of electrode polarization and drift. Carrying out leap frog surveys in closed loops does not provide an absolute check, because the polarization of an electrode pair may change magnitude and polarity from one reading to another. It is helpful to alternate the positions of the leading and following electrodes in leap frogging so as to reduce the cumulative error caused by electrode polarization. In the "total field" configuration, where one electrode is kept fixed during the entire survey and the second electrode is moved, the error at each reading is limited to the maximum value of polarization. When adequate care is taken in the measurements, the self-potential observations can be repeated to within ± 5 mV, even after many years (Parasnis, 1970).

As reported by Corwin and Hoover (1979), self-potential anomalies ranging from about 50 mV to over 2 V have been recorded in over a dozen geothermal areas. No consistent pattern is observed, although the steepest gradients are often associated with faults, fissures and openings which act as conduits for geothermal water. The broad-scale anomalies are often found to be associated with the areas of high heat flow.

Seismic methods

In recent years, there has been much new development in the application of seismic methods in geothermal exploration, and these have been recently reviewed by Iyer (1978). They can be broadly classified into *active* and *passive* methods. Active methods make use of seismic signals generated by an artificial source such as explosives and vibrators, whereas passive methods make use of natural sources such as earthquakes, microseisms and seismic noise.

Active methods

These methods include seismic reflection and refraction surveys, and travel time residual and attenuation studies. Although active seismic methods have been very successfully used in exploration for oil and the theory and field procedures are pretty well established, their use in geothermal exploration has been rather limited.

Refraction profiles have been useful in estimating the depth of the geothermal reservoir in the Reykjanes geothermal field in Iceland (Palmason, 1971). At Reykjanes it is interesting to note that geothermal aquifers are more abundant in deeper, higher-velocity ($V_p \simeq 4.2$ km/s) layers and are not so often located in the shallower, low-velocity material ($V_p \simeq 3.0$ km/s). It has been also discovered that the highest porosity rocks are not necessarily always most productive (Macdonald and Muffler, 1972).

Seismic refraction surveys were found useful in evolving an average crustal model in Long Valley, California (Hill, 1976). Strong attenuation of high-frequency seismic waves was also found to be associated with surface geothermal phenomena in Long Valley, California (Hill, 1976). Evidence for the presence of material with high-velocity and high-attenuation characteristics in the top 1 km layer at The Geysers geothermal field in California have been found by Majer and McEvilly (1979) using an array of closely spaced seismometers. Combs et al. (1976a) have reported results of investigating lateral variation in compression wave velocity and attenuation in the East Mesa geothermal field, California, using explosive charges of from 9 to 90 kg at depths varying from 15 to 60 m. They delineated a low-velocity layer with a horizontal extent of a few kilometers beneath the East Mesa geothermal field. A few other surveys using active seismic methods have been reported in the literature (Iyer, 1978); however, their use in geothermal exploration is quite limited.

Passive methods

Seismic ground noise surveys. A high level of seismic noise is invariably found to be associated with geothermal areas. The seismic noise surveys are carried out using a closely spaced group of seismic stations recording for at least 48 hours (Iyer and Hitchcock, 1976). The process is repeated in different parts of the geothermal area. It is desirable to operate one reference station continuously. Analysis techniques include computation of average noise level in several frequency bands, using carefully selected noise samples, and plotting their spatial variations. Very often power spectra, cross-spectra, azimuth and velocity waves are also calculated. Iyer and Hitchcock (1976) have summarized the results obtained at four geothermal fields in the United States: The Geysers, Imperial Valley and Long Valley in California and Yellowstone National Park in Wyoming. The basic results are:

(1) All four seismic areas have high noise levels in the 1- to 5-Hz frequency band.

(2) At the Yellowstone geyser basins the noise is generated at depth and it is not a surface phenomena.

(3) The cultural noise present at The Geysers and the East Mesa area in the Imperial Valley makes it difficult to interpret the geothermal noise.

(4) The noise anomaly found at Long Valley is partly caused by the amplification of seismic waves by soft alluvial basins.

Iyer (1978) concludes that although tens of noise anomalies have been reported from geothermal areas, attempts at discriminating geothermal noise from cultural noise have not been very successful. More experimental work under varying geological conditions needs to be carried out to develop the seismic noise survey as a useful geothermal exploration tool.

Teleseismic P-wave delays. Seismic waves arriving from earthquakes occurring at far-off distances (larger than 1000 km) are called teleseismic waves. The teleseismic P-wave delay technique, as applied in geothermal areas, involves measuring delays in the relative travel time of teleseismic P-waves that have traversed the geothermal area. Usually an array of seismographs with accurate time keeping is deployed and one of the stations in the array is chosen as the reference station and relative differences in arrival times at the other stations are calculated. These delays are then interpreted in terms of subsurface structure and low-velocity layers. Fig. 5-12 shows the delays observed by Steeples and Iyer (1976) at the Long Valley Caldera, California, and Fig. 5-13 shows a model of subsurface structure interpreted by them along section AA' of Fig. 5-12. It is estimated, according to one of the models, that a volume of about 1000 km^3, with temperatures in excess of about 300°C above normal exists at depths of 10—20 km below the surface. Iyer et al. (1979) have reported

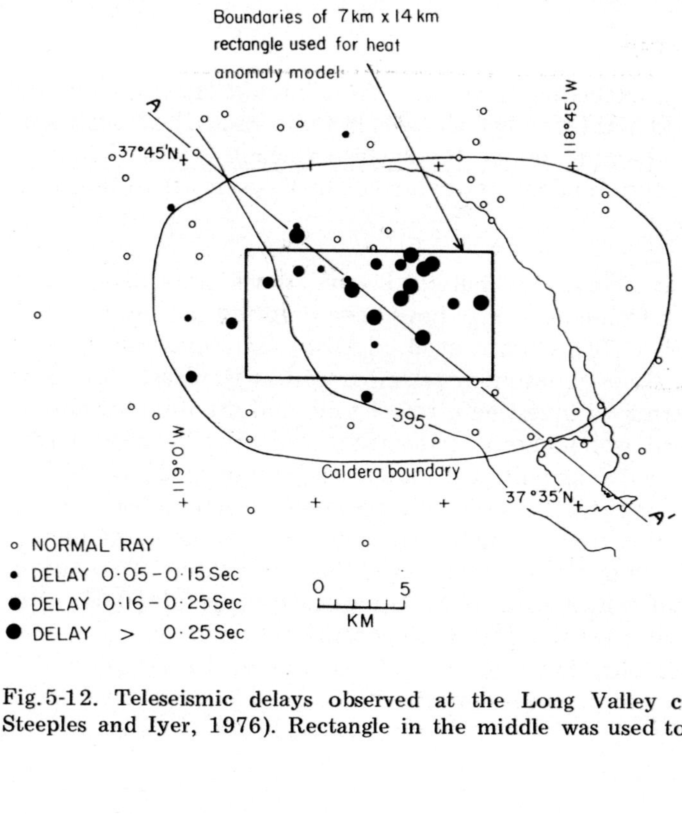

Fig. 5-12. Teleseismic delays observed at the Long Valley caldera, California (from Steeples and Iyer, 1976). Rectangle in the middle was used to estimate heat anomaly.

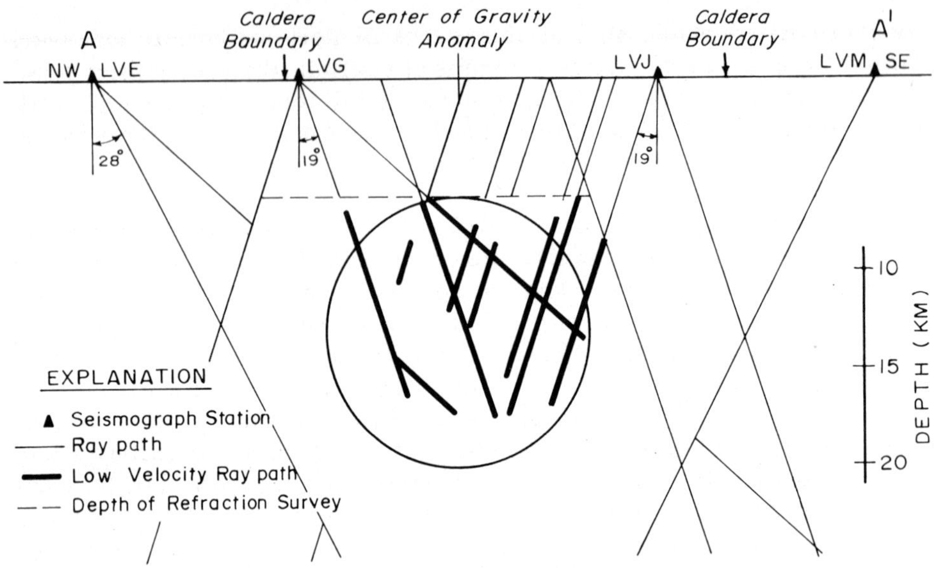

Fig. 5-13. Sectional view of Long Valley caldera along *AA'* of Fig. 5-12. A velocity contrast of 15% along ray paths shown by thick lines, with the ray path length proportional to delays seen along those paths, is represented in the model (from Steeples and Iyer, 1976). For details, see text.

large teleseismic delays, exceeding 1 second, from Mount Hanna, near the Clear Lake volcanic field and from The Geysers in California. They postulate that a molten magma chamber under the surface volcanic rocks of Mount Hanna and a highly fractured steam reservoir at The Geysers are responsible for the observed delays.

P-wave attenuation and delay. Anomalous P-wave travel time delays and attenuations in the high-frequency range have been reported for a number of geothermal fields (Iyer, 1978). Combs et al. (1976b) used an array of nine short-period, high-gain three-component seismographs at the East Mesa geothermal field in California to investigate travel time and attenuation anomalies. Records of several well-located earthquakes from the Brawley earthquake swarm of 1975, with epicentral distances varying from 20 to 50 km were examined. They discovered significant P-wave travel time delays for ray paths passing through the zone of high heat flow. Spectral analyses of the observed seismic waves from the swarm showed that the relative attenuation of body wave amplitudes increased in the frequency range of 10 Hz and higher along the paths through the East Mesa geothermal field.

Young and Ward (1980) have developed techniques to estimate the attenuation of teleseismic P-waves and have discovered a zone of large attenuation coinciding with a zone of large delays at The Geysers in California. Similar results have also been reported for Coso geothermal area (Young and Ward, 1978).

Surface wave dispersion. In contrast to seismic body waves, whose depth of penetration in a given medium is governed by the distance between the source and the observation point, the depth of penetration of surface waves is controlled by wavelength, and the corresponding depth resolution is well established. This property makes surface waves particularly suitable for a three-dimensional estimate of shear wave velocity structures. In spite of the extensive application of seismic methods in petroleum exploration, however, surface wave analysis has played virtually no part in the study of shallow structures. This fact can be traced to the use of explosive sources, by which the surface waves are poorly excited and seldom observed (Dobrin et al., 1954). Geothermal systems, however, are characterized by an abundance of natural seismicity (e.g., Ward, 1972; Hill et al., 1975; Combs and Rotstein, 1976), which produces relatively greater surface wave excitation than explosive sources (Richter, 1958). Since not much is available in the literature on the utilization of surface waves in geothermal areas, this topic is dealt with here in detail.

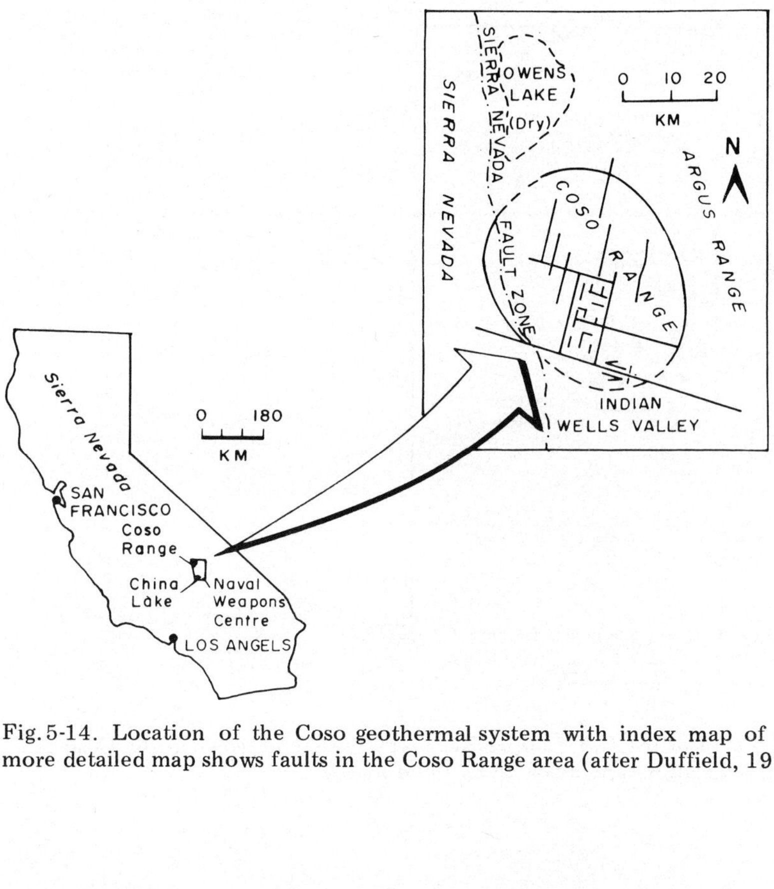

Fig. 5-14. Location of the Coso geothermal system with index map of California. The more detailed map shows faults in the Coso Range area (after Duffield, 1975).

Seismic stations have been installed at a number of geothermal areas around the world. This makes it possible to investigate interstation surface waves and interpret them in terms of shallow crustal structure. We shall briefly examine the results of an investigation carried out at the Coso geothermal fields in California, U.S.A. (Fig. 5-14) which are characterized by the occurrence of Pleistocene volcanic rocks. The area exhibits a high heat flow of up to 15 HFU within the ring-type caldera. A number of seismic stations (Fig. 5-15) are in operation in the area. The inter-station impulse response method of measuring dispersion (Gupta et al., 1977) accurately estimates surface wave dispersion and attenuation for paths between pairs of stations. For two stations which are in line with an epicenter, the surface wave which arrives at the farther station has passed through the nearer station. The signal at the nearer station is considered to be the input to a filter, the output of which is the signal at the farther station. The impulse response of the filter then represents the effect of the path between the stations, independent of the source, the path to the nearer station, and the seismometers when they are identical. Interstation group velocity, phase velocity and attenuation can be calculated from frequency-time analysis (e.g., Dziewonski et al., 1968; Nyman and Landisman, 1977) of the impulse response. The impulse response

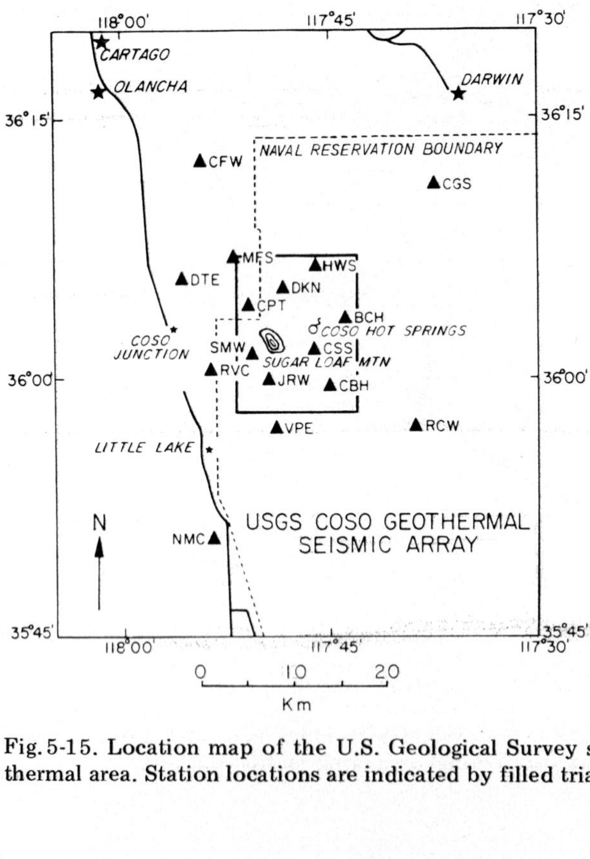

Fig. 5-15. Location map of the U.S. Geological Survey seismic network in the Coso geothermal area. Station locations are indicated by filled triangles.

$x_1(t)$ is the inverse Fourier transform of the Fourier transform of the signal at the farther station, x_2, divided by the Fourier transform of the signal at the nearer station x_1 (Gupta and Nyman, 1977a):

$$x_i(t) = \frac{1}{2\pi} \int_{-\infty}^{\infty} [F_2(\omega)/F_1(\omega)] \exp(i\omega t)\, d\omega \tag{5-22}$$

where:

$$F_j(\omega) = \int_{-\infty}^{\infty} x_j(t) \exp(-i\omega t)\, dt, \qquad j = 1, 2 \tag{5-23}$$

Gupta and Nyman (1977b) have recently applied the inter-station impulse response method to investigate the subsurface velocity structure in the Coso geothermal area, California. One of the highlights of their investigation is the confirmation of a low Poisson's ratio in this area, which was earlier inferred by Combs and Rotstein (1976).

Fig. 5-16 shows seismograms at two stations located at distances of 6.7 and 3.3 km from the epicenter of a shallow earthquake. The epicenter and the seismic stations lie approximately on a straight line. Fig. 5-17 shows the results of the display-equalized frequency-time analysis described earlier. From this, a Rayleigh wave group velocity of about 2.8 km/s is inferred in the period range of 0.07—0.15. This implies that Rayleigh wave velocities to a depth of about 300 m are 2.80 km/s. It is well known that Rayleigh waves propagate with a velocity of approximately nine-tenths that of S-waves. Therefore, the S-wave velocity would be 3.11 km/s to the depths of 200 to 300 meters from the surface. Detailed P-wave velocity data obtained from nine calibration blasts show a constant P-wave velocity of 4.75 km/s. These P- and S-wave velocities infer a Poisson's ratio of 0.13 compared with the 0.25—0.30 normally observed.

The low Poisson's ratio observed for the Coso geothermal field implies that the shallow subsurface is deficient in water content, or more likely the

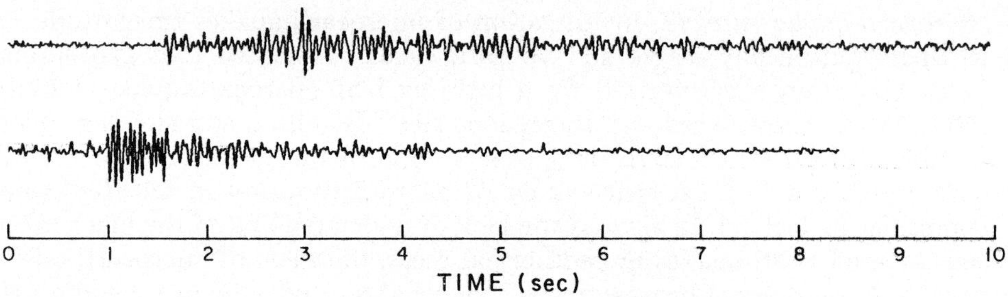

Fig. 5-16. Seismograms at two seismic stations in the Coso geothermal area for a small earthquake. The epicentral distances are 3.3 km (lower seismogram) and 6.7 km, respectively.

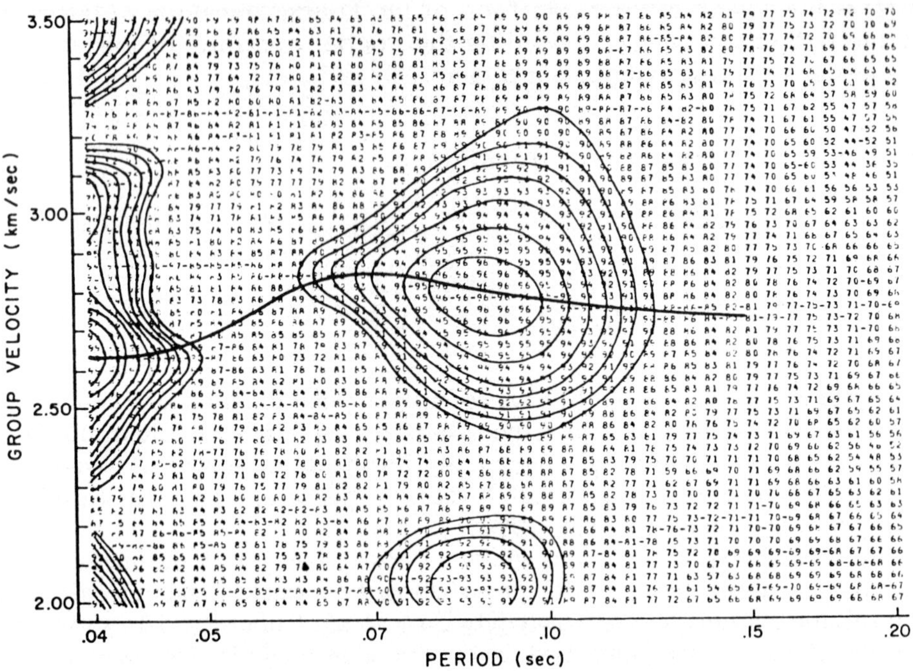

Fig. 5-17. Display of equalized frequency-time analysis of the inter-station impulse response for Coso stations. Inter-station impulse response has been calculated from the seismograms shown in Fig. 5-16. Power in decibels (dB), normalized to a maximum of 99, is contoured at 1-dB intervals. Minus signs designate relative maxima of power as a function of group velocity. Thick line indicates the inferred group velocities.

cracks and void spaces are filled with steam. Similar results from other considerations have been reported by Combs and Rotstein (1976). These data help in inferring that the Coso geothermal field is a vapor-dominated rather than a hot water system.

Microearthquake surveys. Investigation of microearthquakes (magnitude −1 to 3) in tectonically active and volcanic areas has shown that geothermal fields are often characterized by a high level of microearthquake activity (Ward et al., 1969; Ward and Bjornsson, 1971; Hamilton and Muffler, 1972; Combs and Rotstein, 1976, and others). Bjornsson and Einarsson (1974) have reported a very extensive study of microearthquakes in the Reykjanes Peninsular in Iceland. In view of the lack of understanding of the mechanism causing microearthquakes in geothermal areas, the value of microearthquake surveys in geothermal prospecting is somewhat limited. One of the main uses of monitoring and precisely locating microearthquake hypocenters is their application in estimating the depth of fluid circulation in hydrothermal systems, which is difficult to estimate otherwise.

Gravity surveys

Gravity surveys are rather simple and inexpensive. However, good elevation control is necessary and in rugged terrain, a considerable amount of effort is required to incorporate the necessary terrain correction. Therefore, gravity surveys are most useful in plains or areas of smoothly varying terrain. Worden gravimeters are most commonly used. Nettleton (1971) has given a brief description of gravimeters, gravity data acquisition and reduction.

Local gravity anomalies in geothermal fields are generally associated with metamorphism or increase in density of the sediment due to deposition of minerals from the rising plumes of thermal water. Meidav and Rex (1970) report that every known geothermal field in the Imperial Valley, California, is associated with a measurable residual gravity anomaly varying from 2 to 22 mgal. These anomalies are believed to have originated from the combination of:(1) intrusive bodies which may have been the original source of the geothermal anomaly, and (2) precipitation of minerals out of the thermal water at a shallower depth. In New Zealand geothermal areas, the observed positive residual gravity anomalies are considered to be caused by rhyolitic domes and hydrothermal alteration of reservoir rocks (Hochstein and Hunt, 1970; Macdonald and Muffler, 1972).

In regions where other geological and geophysical considerations indicate the presence of economic geothermal reservoirs, gravimetry surveys are useful in providing a clue to identifying localized subareas where geothermally-related processes may have taken place. The reverse is not true. Gravity anomalies alone cannot necessarily be indicative of a geothermal region.

Magnetic surveys

The usefulness of magnetic surveys in geothermal exploration is controversial (Cheng, 1970; Banwell, 1970). In volcanic areas, the magnetization of various rock units is quite different. Large volumes of rocks are altered in hydrothermal areas and this could be best detected through aeromagnetic surveys which are less affected by near-surface rocks. The magnetometers used and method of reduction of data generated by ground magnetic surveys have been discussed in detail by Nettleton (1971).

Magnetic anomalies in the New Zealand geothermal fields have been interpreted as having been caused by conversion of magnetite to pyrite (Studt, 1964). Distinct magnetic anomalies are found to be associated with the high-temperature geothermal fields. Namafjall and Krafla geothermal fields in northern Iceland are good examples (Palmason, 1976). Magnetic ground surveys are also extensively used in the low-temperature fields of Iceland for tracing hidden dykes and faults which often control the flow of geothermal water to the earth's surface.

AIRBORNE SURVEYS

Airborne surveys are known to be a very quick reconnaissance tool. Several efforts in the past have been made to use thermal infra-red imagery, multiband black and white photography, color photography and infra-red color-shift photography for detecting and mapping thermal activity. New geothermal areas have been reported to be discovered in New Zealand (Dickinson, 1973) and Italy (Hodder et al., 1973) with the help of aerial infra-red surveys. Dupart and Omnes (1976) recommend that high-sensitivity aeromagnetic surveys should also be considered in geothermal exploration programs. Marsh et al. (1976) report the use of satellite data for geothermal exploration.

It should be borne in mind that both the photographic and thermal infra-red sensors can make measurements only to depths of a few microns from the surface. This limitation is due to the normal opacity of soil at these wavelengths. Therefore, only surface effects can be detected. Photographic sensors (visible and infra-red) are usable in identifying hydrothermal alteration associated with thermal seep or fumaroles. Moreover, photographic sensors can operate only during the time when the reflection of incident sunlight to detect thermal manifestations is available.

In the following the fundamentals of infra-red radiation and atmospheric transmission windows and the application of airborne surveys in geothermal prospecting are briefly discussed.

Infra-red radiation and atmospheric transmission windows

The electromagnetic spectrum between the wavelengths of 0.7 and 1000 μm (7×10^{-5} and 10^{-1} cm) is known as the infra-red radiation. The lower wavelength limit coincides with the upper limit of visible radiation. Radiations with wavelengths in excess of 1000 μm are called the microwave spectrum. Both these limits, lower and upper, are arbitrary and no change in characteristics occurs as they are passed. Conventionally, the 0.7- to 1.5-μm region is called the near infra-red region, from 1.5 to 20 μm the intermediate infra-red region and from 20 to 1000 μm the far infra-red region.

All bodies above absolute zero in temperature emit energy as electromagnetic radiation. In the infra-red region this radiation is governed by the Stefan-Boltzmann law, as discussed in Chapter 3. Emissivity, the ratio of the energy emitted by a material to that emitted by a black body, for the earth materials ranges from 0.70 to 0.98 compared to a black body emissivity of 1.0.

A large portion of the infra-red radiation is absorbed in the atmosphere and consequently certain wavelengths, for which absorption is low, can only be detected by airborne sensors. Fig. 5-18 shows the atmospheric transmission at different wavelengths of the infra-red spectrum.

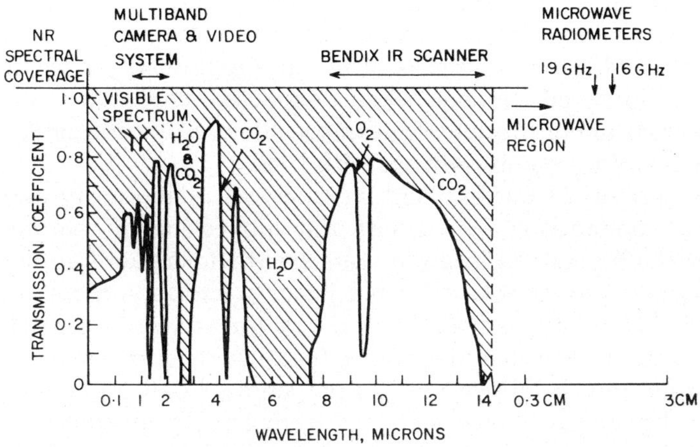

Fig. 5-18. Atmospheric transmission and sensor coverage (from Hodder, 1970). For details, see text.

Infra-red aerial surveys

Aerial infra-red surveys furnish useful information that can be related to the heat discharge at the earth's surface. Up to recently (e.g. Banwell, 1973), it was thought that only areas with heat discharges of the order of 300 HFU could be identified with reasonable certainty. This is about 200 times the average heat discharge — and such areas are not common. However, after removing non-geothermal effects, it appears possible to enhance the applicability of infra-red surveys (Del Grande, 1976) for identifying potential geothermal areas characterized by much smaller heat flow values.

Several airborne surveys with varying degrees of success in the reconnaissance of geothermal resources have been reported in the literature. Hochstein and Dickinson (1970) have used infra-red scanning equipment operating in the 4.5- to 5.5-μm range in the Karapiti area, near Wairakei, New Zealand, to determine whether infra-red surveys could monitor changes in the boundaries of discharge areas over a period of a few years. They conducted a trial run at noon on an overcast day over a selected strip at an altitude of about 1000 m. A comparison with panchromatic and reflective infra-red photographs and near-surface temperature measurements revealed that the infra-red scanner picture displayed all the essential thermal features in relatively great detail.

Palmason et al. (1970) have reported results of aerial infra-red surveys at the Reykjanes and Torfajokull thermal areas in Iceland. The surveys were conducted with an infra-red scanner system in the 4.5- to 5.5-μm wavelength band. To interpret features depicted on infra-red imagery and to relate zones of high heat flux to tectonic structure, supplementary geological studies were made. These included preparation of a shallow ground temperature map at a depth of 50 cm. Aerial infra-red surveys were found useful in outlining the surface thermal patterns and in relating them to possible geological structures controlling the upflow of hot water. Palmason et al. (1970) conclude that in addition to their use in the preliminary study of high-temperature areas, infra-red surveys, when conducted at regular intervals over geothermal areas under production, can provide useful information on changes occurring in surface manifestations with time.

Hodder (1970) has presented an excellent summary of a laboratory and field research program undertaken to evaluate the application of remote-sensing techniques in geothermal exploration. Multiband photography in the visible and near infra-red band, passive infra-red imagery and radiometry in the 8- to 14-μm band, as well as passive microwave radiometry at 16 and 19 GHz (1.875 cm and 1.58 cm wavelengths, respectively) were used. The field experiments were carried out in known geothermal areas of California with various geological backgrounds. The effects of varying flight altitudes on thermal anomaly signatures were also investigated. Fig. 5-19 shows the relationship between flight altitude and the corresponding applications. The major result of this research program was the identification of temperature anomalies in the microwave and the Rayleigh-Jeans region of spectrum in addition to the usually used infra-red region.

Dickinson (1976) has reported aerial infra-red surveys in the 4.5- to 5.5-

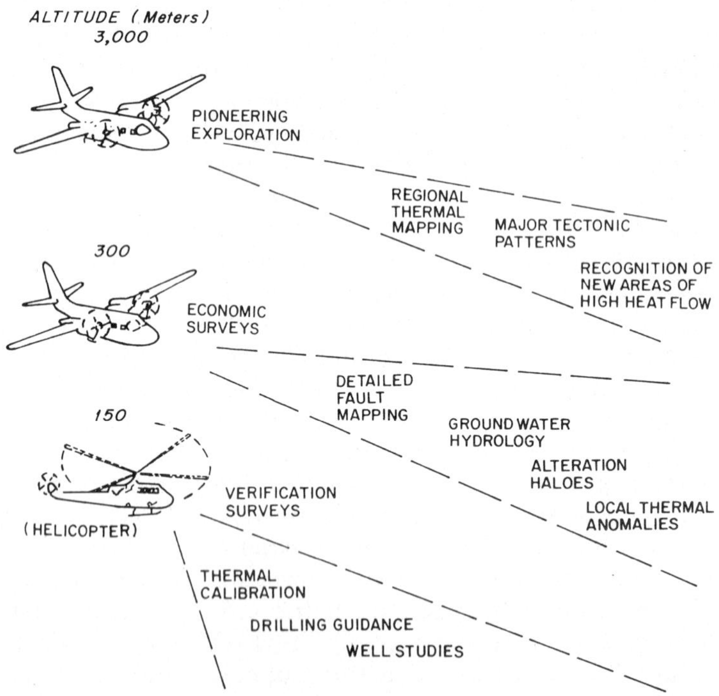

Fig. 5-19. Flight altitudes and corresponding coverage (from Hodder, 1970).

μm band, flown over urban, industrial and undeveloped land in the Tauhara and other geothermal fields in New Zealand. Maps have been prepared from the thermograms dividing the surface temperatures into three categories. These are at ambient temperature, 1—3°C above ambient, and more than 3°C above ambient. The 3°C above ambient temperature contour coincides with the boundary between regions of conductive and convective heat flow, as determined from surface measurements. Dickinson (1976) succeeded in identifying some previously unknown areas of warm ground and in identifying and removing thermal anomalies of non-geothermal origin. He proposed repeating similar surveys at intervals of approximately three years to monitor any changes in hydrothermal activity.

Del Grande (1976) has reported a new aerial reconnaissance method under development for identifying and evaluating potentially valuable geothermal resource areas. A geothermal energy multiband (GEM) detection system has been designed. This system is expected to resolve 0.05—0.5°C temperature enhancements for areas larger than 1 km^2 by ratioing narrow infra-red spectral bands at 2.2, 3.5, 3.9, 4.8 and 13.2 μm. These signal ratios have the advantage of being insensitive to surface emissivity for natural terrains. The signal ratios vary as a high power of the absolute surface temperature, and they avoid the 6- to 13-μm region which has interpretive uncertainties. Del Grande (1976) recommends undertaking four pre-dawn surveys under varied climatic conditions to provide redundant data for distinguishing geothermal effects from climatic thermal noise. After successful development, the GEM system is expected to have a 10-times better sensitivity than the conventional airborne infra-red surveys.

Banwell (1970), in his review of geophysical techniques in geothermal exploration, concludes that infra-red surveying is useful in detecting and mapping relatively strong geothermal anomalies under favorable conditions using optimum equipment and data-handling facilities. However, its application is constrained by various random noise effects and other uncertainties. Therefore, careful considerations of various factors is necessary before undertaking airborne infra-red surveys for geothermal exploration.

Aeromagnetic investigations

Several airborne magnetic surveys have been made over geothermal areas. Various methods of interpreting magnetic anomalies in geothermal areas have been suggested. Here some salient features of airborne magnetic methods are given.

Detection of geological features such as faults, horsts, and grabens is important in geothermal exploration. Under favorable circumstances, these features can be successfully picked up by high-precision aeromagnetic surveys. Dupart and Omnes (1976) have listed the technical requirements for such surveys. These include: (1) constant barometric flight level, (2) small average ground clearance, (3) tight grid of profiles, (4) sensitivity of 0.01 or

0.02 gamma, ensuring a tenth of a gamma as a usable measurement, (5) digital recording at an interval of half a second, (6) accurate navigation, and (7) removal of diurnal variations.

Aeromagnetic surveys are quite expensive. Dupart and Omnes (1976) estimated the cost then to be U.S. $125 per square kilometer. High-precision aeromagnetic surveys have been successful in estimating the sedimentary cover overlying crystalline basement quite accurately. Deguen et al. (1974) report that surveys carried out over a shallow basin in Canada were very successful and the depth of basement estimated from aeromagnetic surveys was found to be within 10% of the depth obtained through drilling at eight drilling sites out of a total of twelve drill holes. McEuen (1970), for Imperial Valley in California, suggested that the magnetic intensity deduced from airborne survey is solely dependent on depth of basement. He found it possible to obtain useful information regarding temperature distribution at depth by comparing the apparent resistivity values obtained from long-spacing resistivity profiles with magnetic intensity values obtained from similar profiles generated from an airborne magnetic map. Another interesting example of the application of airborne magnetometry to geothermal exploration has been presented by De la Funte and Summer (1974) for the Colorado River delta area, Baja California, Mexico.

Utilization of satellite data

The U.S. National Oceanographic and Atmospheric Administration (NOAA) meteorological satellite is the most useful satellite equipped with thermal infra-red sensors and with reasonable ground resolution for geological and geophysical applications. Marsh et al. (1976) have investigated the applicability of the NOAA Very High Resolution Radiometer (VHRR) to geothermal studies in the continental U.S. The VHRR senses energy in the visible spectrum at 0.6—0.7 μm and in the thermal infra-red at 10.5—11.5 μm. The satellite covers the continental United States twice a day with a ground spatial resolution of approximately 1 km. Marsh et al. (1976) report that a preliminary analysis of the computer-derived images of the geothermal areas of Yellowstone National Park, Wyoming, indicated that the delineation of some geothermal areas is possible. For a better utilization of data, problems of adequate thermal and spatial resolution and appropriate local pass times need to be resolved. Additionally, at satellite altitudes, problems of atmospheric attenuation are compounded. Marsh et al. (1976) have concluded that until more technologically advanced thermal-sensing systems, such as Earth Resource Technological Satellite-C (ERTS-C), are made available to the scientific community, the NOAA system provides an opportune research instrument of sufficient capability for geothermal studies.

CONCLUDING REMARKS

As pointed out by McNitt (1973), it is advisable not to depend on a single method or indication alone in the selection of geothermal prospect areas, whether it be geological, hydrogeological, geochemical or geophysical. To answer specific questions, individual surveys should be designed and the results obtained should be verified from other available data or surveys. Well locations for drilling should be selected after all surveys have been completed and internal ambiguities and inconsistencies have been resolved as far as possible.

Chapter 6

ASSESSMENT AND EXPLOITATION

DRILLING TECHNOLOGY

Drilling is one of the areas where geothermal resource development has gained considerably from the expertise of the oil and gas industry. Drilling for geothermal energy is quite similar to rotary drilling for oil and gas. The main differences are due to the high temperatures associated with geothermal wells which affect the circulation system and the cementing procedures as well as the design of the drill string and casing. In addition, relatively much

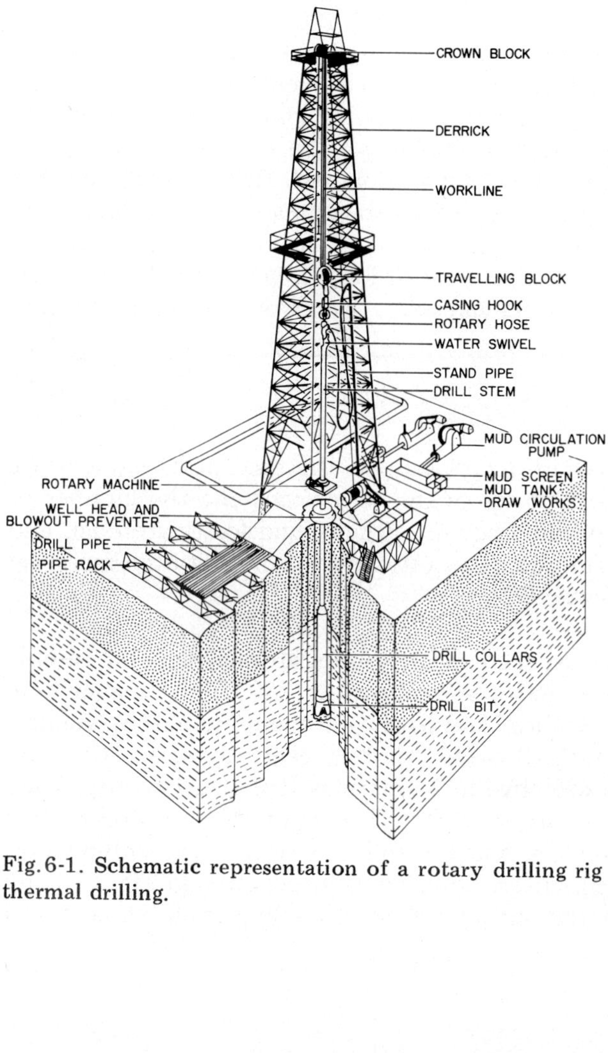

Fig. 6-1. Schematic representation of a rotary drilling rig and the peripherals used in geothermal drilling.

harder fractured volcanic rocks are encountered in most geothermal areas. Geothermal steam is at some places encountered at shallow depths of 50—200 m, while at others drilling to depths in excess of 3000 m has been necessary. During the last three decades, drilling technology has witnessed a great advancement. In 1947 only about a dozen holes deeper than 5000 m existed, but by 1972 the number of such holes reached the 500 mark. At the same time, through superior metallurgy, the bit life increased by twenty-fold. The hoisting capacity of deep-well rigs has also increased 4—5 fold through the use of heavier and more efficient equipment. Besides overall improvements being made in different aspects of rotary drilling, research is at present being conducted to develop novel drilling techniques (Maurer, 1976). The four basic mechanisms being tried are: melting and vaporization, thermal spalling, chemical reactions and mechanically induced stresses.

In conventional rotary drilling for geothermal energy, a normal oil-well rig such as the one shown in Fig. 6-1 is used. A typical rig to drill a 1000- to 2000-m deep hole and to provide adequate hoisting capabilities would have a 400- to 500-hp (horse power) engine to operate a 300-hp draw-works, and an independent 500-hp mud pump to supply the hydraulics needed to drill efficiently; the mast and substructure would have a capacity of over 100 tons to provide an adequate margin of safety in running the liner. A drillstring of $3\frac{1}{2}''$ drill pipe and $6''$ drill collar would be sufficient. To facilitate mobility, all the components would preferably be trailer mounted. Mud is usually used as the drilling fluid. However, the use of air instead of mud makes drilling much faster and cheaper and has drawn increasing attention during the recent years. One major obstacle in applying air drilling in geothermal areas is its unsuitability in formations bearing excessive water or in formations that tend to slough.

Drilling with mud

The time required for drilling depends upon the ease of access and the topography of the site, the hardness of the formations and the number of meters to be drilled in the productive zone where mud losses are high, and on the amount of casing required. The estimates of drilling time made by Matsuo (1973) are given in Table 6-1.

Drilling mud. Clay-based muds, such as bentonite, are the most commonly used in geothermal drilling at low temperatures (~150°C). At higher temperatures, ordinary clay-based muds tend to gel and therefore chrome-lignite/chrome-lignosulphonate (CL-CLS)-based mud is generally used. CL-CLS-based muds are particularly effective in formations that cave in easily. They can withstand temperatures of up to 250°C. For more efficient drilling it is desirable to keep the temperature and the solid content of the drilling mud as low as possible. The temperature is often reduced through 10—15°C by installing cooling towers at the surface. Further cooling could be achieved

TABLE 6-1

Estimates of drilling time in geothermal areas under favorable conditions (Matsuo, 1973)

Depth (m)	Actual drilling (days)	Finishing and testing (days)
500	15–30	10
1000	25–45	10
1500	35–55	10
2000	50–70	10

with the help of large fans. The solid content of the mud is sometimes mechanically reduced by a large, high-speed shaker, a desander, and a desilter.

Drilling bits. Roller bits with hardened-steel teeth or tungsten-carbide inserts are commonly used in geothermal drilling (Fig. 6-2A). Since the steel used in roller bits is drawn at temperatures of about 200–250°C, these bits lose much of their strength when operated at temperatures in excess of 250°C. This causes rapid failure of bearings and steel teeth as well as loss of inserts with the insert bits. Expensive roller bits are provided with sealed lubrication systems with rotating rubber seals to hold the grease in the bearings. Since these rubber seals have a temperature limitation of about 200°C, improved seals and improved high-temperature lubricants are required in high-temperature geothermal drilling. Diamond drills (Fig. 6-2B) can drill at temperatures in excess of 500°C. However, since their drilling rate is much slower compared to roller bits, they do not provide a very acceptable solution to the problem of high-temperature drilling.

Casing. To obtain a sustained flow of steam from a reservoir, it is necessary to choose an appropriate diameter for the production well. Additionally, it is necessary to provide adequate casing at correct depths to prevent hot

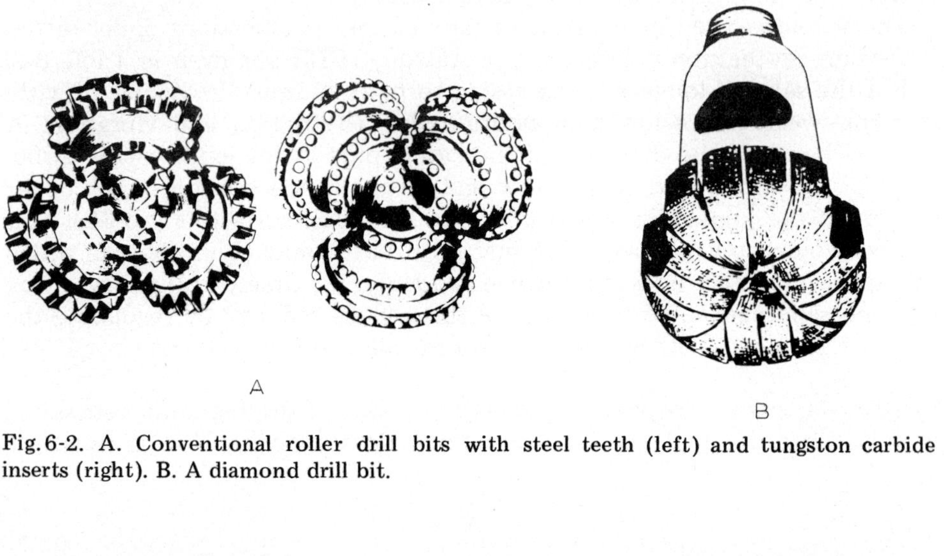

Fig. 6-2. A. Conventional roller drill bits with steel teeth (left) and tungston carbide inserts (right). B. A diamond drill bit.

TABLE 6-2

Recommended borehole diameters and casing specifications for different steam production volumes (Matsuo, 1973)

(1) Steam volume: 10—25 tons/hour		
17″ open hole	13-3/8″ surface casing	fully cemented
12-1/4″ open hole	9-5/8″ intermediate casing	fully cemented
8-5/8″ open hole	7″ production casing	fully cemented
6-1/4″ open hole	4-1/2″ slotted liner	
(2) Steam volume: 25—50 tons/hour		
18″ open hole	16″ surface casing	fully cemented
14-3/4″ open hole	11-3/4″ intermediate casing	fully cemented
10-5/8″ open hole	8-5/8″ production casing	fully cemented
7-5/8″ open hole	6-5/8″ slotted liner (outer diameter of coupling skimmed by 1/16″)	
(3) Steam volume: 50—80 tons/hour		
22″ open hole	18″ surface casing	fully cemented
17″ open hole	13-3/8″ intermediate casing	fully cemented
12-1/4″ open hole	9-5/8″ production casing	fully cemented
8-5/8″ open hole	7″ slotted liner	

water from higher formations from entering into the well. When the pH value of the hot water produced with the steam is low, it tends to corrode the casing and reduce the life span of the well.

Slotted liners are recommended when the wall of the hole tends to slough and caving in of the hole is suspected. Sloughing can be recognized by the presence of sloughed-off particles among the drill cuttings when passing through the production zone. If the hole wall collapses, it would take much longer to re-drill when an ordinary liner is used.

The recommended hole diameters and casing specifications under different steam production volumes (after Matsuo, 1973) are given in Table 6-2.

For the sake of longevity, the casing to be used must be capable of withstanding wear, corrosion and attrition due to friction and vibration. A sudden change in the diameter of the casing pipe at the joints causes turbulence in the high-speed steam flow and results in the erosion of the upper corners of the joints (Fig. 6-3A) as well as the inside surface (Fig. 6-3B). This can be avoided by the use of internal flush butt joints (Fig. 6-3C), in which the inside diameter does not change suddenly. At times, casing is worn by the fine sand carried with the steam; this can be reduced by regulating the speed of the steam with a flow regulator installed at the wellhead.

Circulation losses. To prevent circulation losses of drilling mud, cementing is sometimes done at the horizons where losses occur. This is usually a

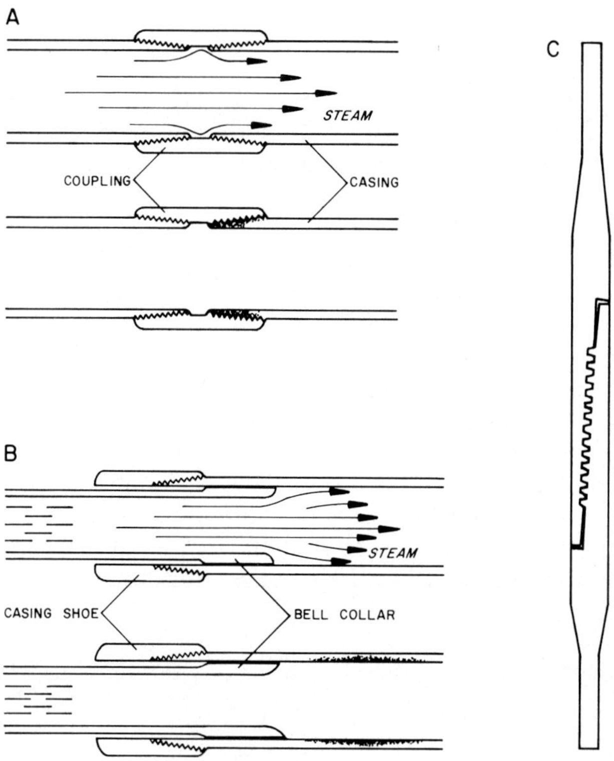

Fig. 6-3. Damage to casing due to variation in the inside diameter at the joints, and its prevention. A. Turbulence in high-speed steam flow caused at the joint (top), and the consequent damage at the corner (bottom) shown by shading. B. Same as A for a bell collar. C. Flush butt joint which helps in preventing such damage.

permanent measure against such losses. At times the formations are plugged with materials such as nutshell powder, cottonseed hulls, cellophane, and fibre scrap. In the course of time these materials carbonise and block the fractured zones.

Cementing. Proper cementing of geothermal wells requires that the cement slurry should rise uniformly and continuously from the casing shoe to the ground level. If it is envisaged that the formation cannot be adequately cemented while drilling, or that it will be difficult to cement the entire height of the hole in one stage, multistage cementing should be planned from the beginning. For proper cementing, the formation of thick filter cakes should be prevented. When thick cakes cannot be prevented, they may be scraped using scratchers. Sufficient clearance between the casing and the wall of the hole should be insured in order for the void to be uniformly filled with cement. If the annular void from the well head to the bottom has not been properly filled with cement, it is usually preferable to provide a rigid support to the head section of the casing instead of attempting to re-inject cement.

Ordinary cement is adequate for temperatures up to 150°C, but to resist higher temperatures, silica is often mixed with it. In geothermal wells where steam is accompanied by low-pH hot water, it is necessary to use acid-resistant cement.

Drilling with air

Rotary drilling with air is quite similar to drilling with mud, but with air replacing the mud. The technique has been tried in many countries for several years and has been particularly successful at The Geysers geothermal field in California. Two important advantages of air drilling over mud drilling are: (1) the production zone is not damaged by the circulating mud, and (2) compared to mud drilling, the drilling speed is 3 to 4 times faster and the bit life is 2 to 4 times longer. However, air drilling is not suitable in formations bearing excessive water and in those that tend to slough. Under suitable sub-surface conditions it is not unusual to drill a hole employing both air and mud drilling.

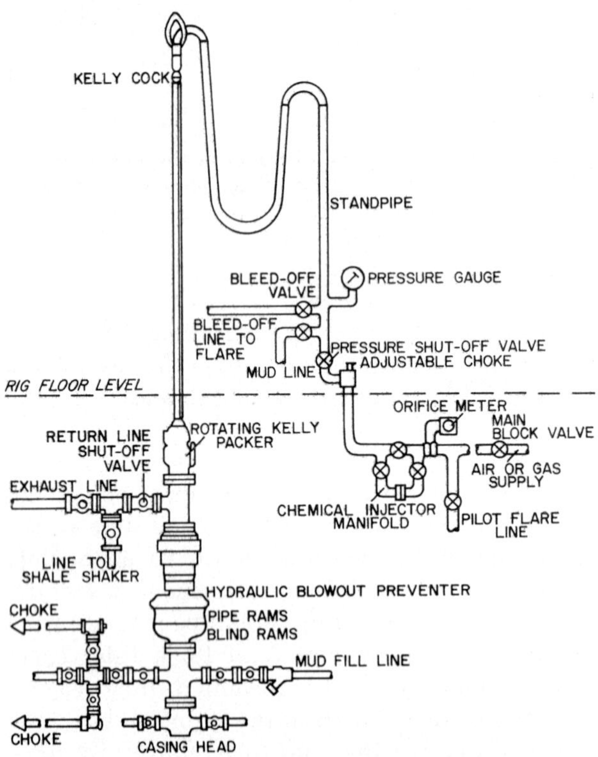

Fig. 6-4. Schematic representation of the air-drilling setup and the peripherals.

Equipment. The equipment used in air drilling is quite similar to that used in mud drilling. Fig.6-4 shows the former process diagrammatically. The mud pump is replaced by an air compressor and the booster compressor. The air compressor provides an air velocity of 600—1500 m/minute in the annular section between the drill pipe and the hole, and the booster compressor provides the additional pressure necessary to blow out the water accumulating in the hole.

The space between the wellhead and the drill pipe is closed with a rotary packer while lowering and lifting the drill pipe. This prevents the air from flowing back to the rotary table and the floor of the derrick. The packer is made of circular strip rubber. It can withstand temperatures of the order of 250°C and may need as frequent replacement as the bit. An inside blow-out preventer is attached to the lower part of the drill pipe. This prevents the steam from flowing out of the drill pipe during raising and lowering. A discharge pipe is used for blowing off air, dust and cuttings. The area of the discharge pipe must be equal to or smaller than that of the annular space between the casing and drill pipe to stop the cuttings from accumulating at the drill collar. Sometimes the discharge pipes are provided with a water sprayer to ease the removal of cuttings and the dust. Compared to the bits used in drilling with mud, the air-drilling bits have enlarged nozzles and air-cooled bearings.

Prevention of water intrusion. The efficiency of air drilling is rapidly reduced with an increase of the amount of water intruding into the well from formations. When the water seepage is very small (e.g. 500 liters or less per hour), it can be successfully stopped by injecting finely ground silica gel and calcium stearate through a pump into the air stream. The quantities injected may be of the order of 2% by weight of the quantity of the cuttings ejected from the well. Calcium stearate forms a coating which repels water from the cuttings and silica gel is helpful in reducing the torque on the drill pipe. When the water seepage is of the order of 500—10,000 liters per hour, a foaming agent such as lithium stearate is used. In case of excessive water seepage (over 10,000 liters per hour), some suitable material such as plastic or cement is used for sealing off the water-bearing horizons. In places silicon tetrafluoride gas has also been used effectively for sealing the formations. When using the gas to prevent seepage, the upper and lower parts of the horizon requiring sealing are sealed with packers and gas is injected in the sealed-off area. On contact with water, a precipitate is formed which blocks the pores of the formation. When water seepage problems are unsurmountable, it becomes necessary to switch over to conventional mud drilling.

Well spacing

In spite of numerous studies, no definite criterion for well spacing in geothermal fields has been formulated. Even when a good cap rock exists, it is difficult to determine the contribution of different underground faults and fissures to the steam produced by a well. Consequently, it cannot be determined how the overall production is going to be affected by siting a well at a certain distance from an existing well. However, in shallow steam fields (500—2000 m deep), a well spacing of 100—300 m is in vogue. Large distances are usually kept to avoid the possibility of pressure interaction between adjacent wells which could lead to a reduction in discharge. James (1976a) has pointed out that geothermal field development costs can be reduced by decreasing the distance between production wells and thus curtailing the length of steam transmission lines, which are quite expensive. He suggests that for geothermal systems where the flow is through fissured formations and the crack permeability is good, the spacing could be somewhat less and successful wells with fairly large discharges could be spaced at distances of the order of 50 m without interaction effects.

Safety measures

Safety measures to be undertaken and their relative importance to one another vary from one location to another. However, in the following a few common problems and their remedies are listed.

(1) Gas detectors and masks should be readily available. Gas emitted by a well may be poisonous and could produce giddiness and eye injury. If emitted in a high concentration, it could even become lethal.

(2) An adequate water supply should be available at the drilling site. At times it takes a long time for the preventer to control the steam gushing out during a drilling operation. Water is useful in averting emergencies. Also, the functioning of the preventers should be checked regularly.

(3) The blow-out preventer is held in position by the casing. To stop the steam from gushing out when the preventer is closed, the space around the casing must be perfectly sealed with cement. The casing shoe must be placed and held firmly in a competent formation and the void outside it must be uniformly and completely filled with cement.

(4) Sometimes it becomes difficult for workers to reach the derrick escape ladder in the event that hot water and steam start gushing out. Hence, it is essential to provide an escape cable and seat.

Repairing of wells

With the passage of time, the pressure in a reservoir declines and consequently the production of steam at the same pressure cannot be sustained with the original casing diameter. It then becomes necessary to insert a smaller-size casing in the well to correspond with the available pressure. The production of a well could also decline or stop due to casing failure, collapses and fractures. The exact procedures followed to clean the well and restore production differ from one location to another. As an example, the following is the procedure adopted at the Cerro Prieto geothermal field in Mexico

(Dominguez and Vital, 1976):

(1) The wells were permitted to cool. The cooling was ascertained by taking several temperature readings.

(2) Pressure, temperature and caliper logs were carried out.

(3) After determining the obstruction, lead impressions were taken to determine its type and size.

(4) Sand plugs were eliminated by mud circulation. The depth of fracture was determined through the flow of sand. Blocked portions of the hole were cleaned by milling to permit the lowering of new production casing and other tools.

(5) Cement grout was injected through the fractures and milled windows to obtain good seals. The seals were checked by hydrostatic tests.

(6) The windows and fractures were covered using $7\frac{5}{8}''$ and $5\frac{1}{2}''$ diameter casings. The lower end of the casing was placed as close as possible to the upper end of the liner and the casings were cemented along their entire length using modified cement. At the end of these operations the plugs, couplings, shoes and so on were scoured until the well was clean.

Novel drilling techniques

It is becoming increasingly necessary to develop faster drilling techniques at elevated temperatures and in deeper holes. In the following some of the concepts being used in developing new methods to meet these requirements are briefly introduced.

The four basic mechanisms being researched are: (1) chemical reaction, (2) melting and vaporization, (3) thermal spalling, and (4) mechanically induced stresses.

Chemical drills. This group of drills employs highly active chemicals to penetrate the rocks. In a typical chemical drill (Ledgerwood, 1960) fluorine is jetted into the rock with the help of a compressed gas chamber. The mechanism is shown schematically in Fig.6-5A. The high cost of the chemicals used is a major factor in preventing large-scale use of this technology.

Melting and vaporization. Devices such as lasers, electron beams, plasmas and electric arcs are capable of heating rocks to temperatures high enough to melt them. Such drills are particularly suitable for drilling in extremely hot reservoirs which cannot be drilled by conventional rotary drills. Fig.6-5B shows a laser drill schematically.

Thermal spalling. When silicate rocks are heated to the 350—550°C temperature range, thermal stresses cause flaking. The jet-piercing drill is one of the tools which uses this principle and it is widely used. This drill burns fuel oil

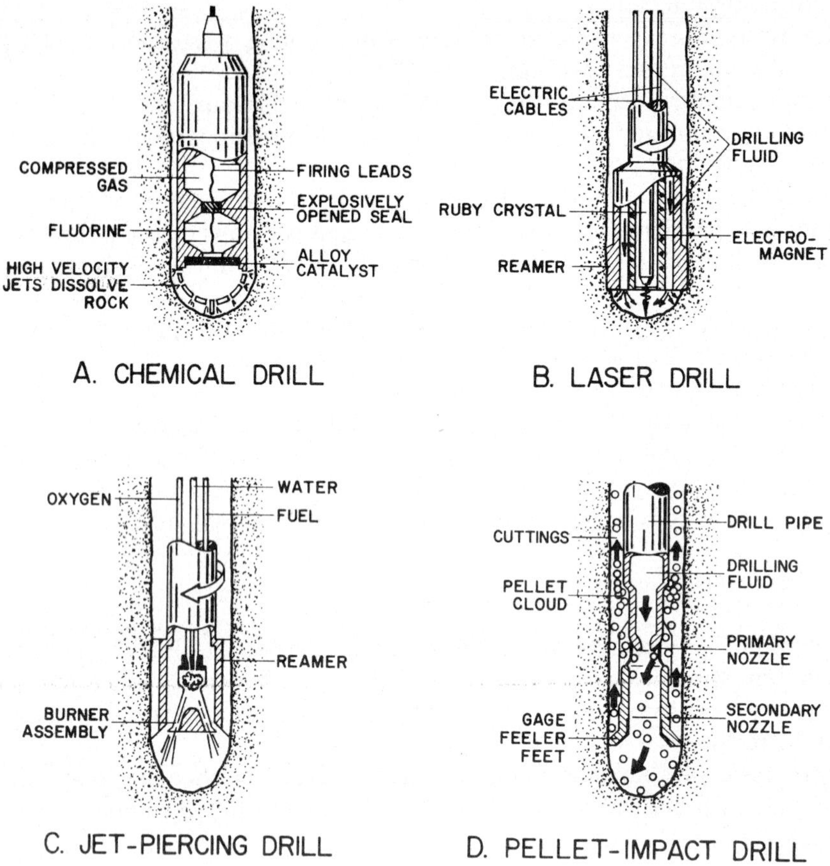

Fig. 6-5. A few novel drilling techniques (from Maurer, 1976). For details, see text.

and oxygen to produce a temperature of 2400°C (Fig. 6-5C). Such drills are capable of achieving drilling rates in excess of 10 m/hr.

Mechanical drills. These drills use turbines, explosives, pellet impact, implosion and many other similar devices to generate mechanical stresses to remove the rock. The pellet impact drill shown schematically in Fig. 6-5D uses nozzle-directed steel pellets to break the rock.

RESERVOIR PHYSICS AND ENGINEERING

Reservoir physics and engineering deal with making the necessary physical measurements, modelling and laboratory studies for estimating subsurface reservoir conditions. The measurements usually carried out in the reservoirs pertain to their size, temperature, pressure, permeability and fluid composi-

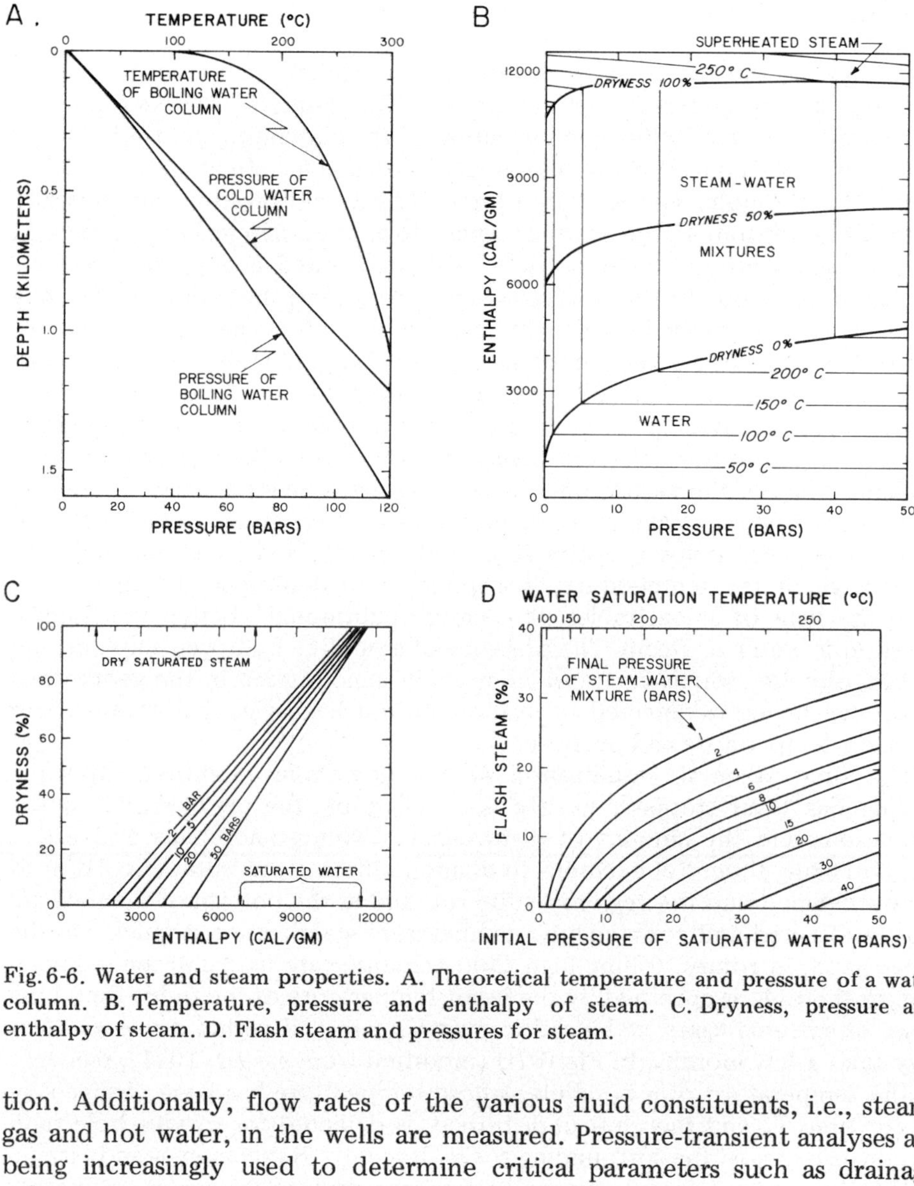

Fig. 6-6. Water and steam properties. A. Theoretical temperature and pressure of a water column. B. Temperature, pressure and enthalpy of steam. C. Dryness, pressure and enthalpy of steam. D. Flash steam and pressures for steam.

tion. Additionally, flow rates of the various fluid constituents, i.e., steam, gas and hot water, in the wells are measured. Pressure-transient analyses are being increasingly used to determine critical parameters such as drainage volume, porosity, permeability and mean formation pressure. Mathematical modelling has been helpful in solving two-phase flow, convection patterns and other problems concerning dynamics of the reservoir under certain simplified assumptions.

The properties of steam and water under varying conditions are known quite accurately and are published in steam tables. These, and other derived properties, such as flash and dryness, are extensively used in geothermal well measurements. Some of these properties are shown in Fig. 6-6.

Subsurface well and reservoir conditions

Because of the disturbances resulting from drilling, it is very difficult to measure the natural reservoir conditions. The reservoir temperatures are temporarily lowered by the cooling action of the circulating mud (Fig. 6-7A). The mass permeability of the reservoir is permanently affected by the introduction of the hole. The measured temperatures and pressures are closer to reservoir conditions in the sections where formation permeability is high. In the uncased sections or the sections with perforated casing, the pressures measured are close to the reservoir values. However, the temperatures may still vary considerably in the sections with fluid flow between permeable zones. The subsurface temperatures can vary unpredictably and inversions are often observed (Fig. 6-7B). Fig. 6-7C depicts some field observations. Hence it is not advisable to interpolate or extrapolate the temperatures. Unlike the temperatures, pressure inversions are never observed and pressure gradient changes drastically only when there is a phase change. Therefore, interpolation and extrapolation of pressures is often done. Measurement of pressure gradient helps in estimating fluid density, and when coupled with knowledge of the temperature, it is possible to determine the fluid type. With the help of steam tables, it is easily determined whether the fluid in question is water or steam. The presence of gases like hydrogen sulphide and carbon dioxide, which are most commonly encountered in the geothermal fields, can be easily detected since they are much heavier than steam under the same temperature and pressure.

The time taken in stabilization of the downhole condition, after the drilling has been stopped, mostly depends upon the permeability of the formations and the amount of disturbance. Drilling particulars also effect the stabilizing time. For example, in diamond drilling the water circulation is relatively small and the regions at the top and at the bottom of the hole are mostly effected and consequently temperature stabilization is quick. On the other hand, in rotary drilling high fluid velocities are used and the temperature in the hole is governed by the input temperature of the fluids for a long time. In extreme cases, stabilization may be achieved within a few hours or may take a few months. In Fig. 6-7D (modified from Jaegar, 1971), recovery in the temperature of a borehole, whose temperature has been changed by θ_0 for time T_0 and then left undisturbed, is depicted for varying hole radii and diffusivities of the surrounding rocks. Recently Kappelmeyer and Haenel (1974) have given useful dimensionless empirical relations for estimating stabilizing time under varying borehole conditions.

Flow measurements, involving physical and chemical methods, are employed to identify and measure the rate of discharge of various constituents being produced at a well. The geothermal fluids — water and steam — are at times contaminated with various other gases like H_2S and CO_2 and solid impurities like sand and salt. The energy content of the fluid arriving at the wellhead is practically the same as that when it enters the borehole. Water or mixtures of water and steam occurring at high temperatures boil during up-

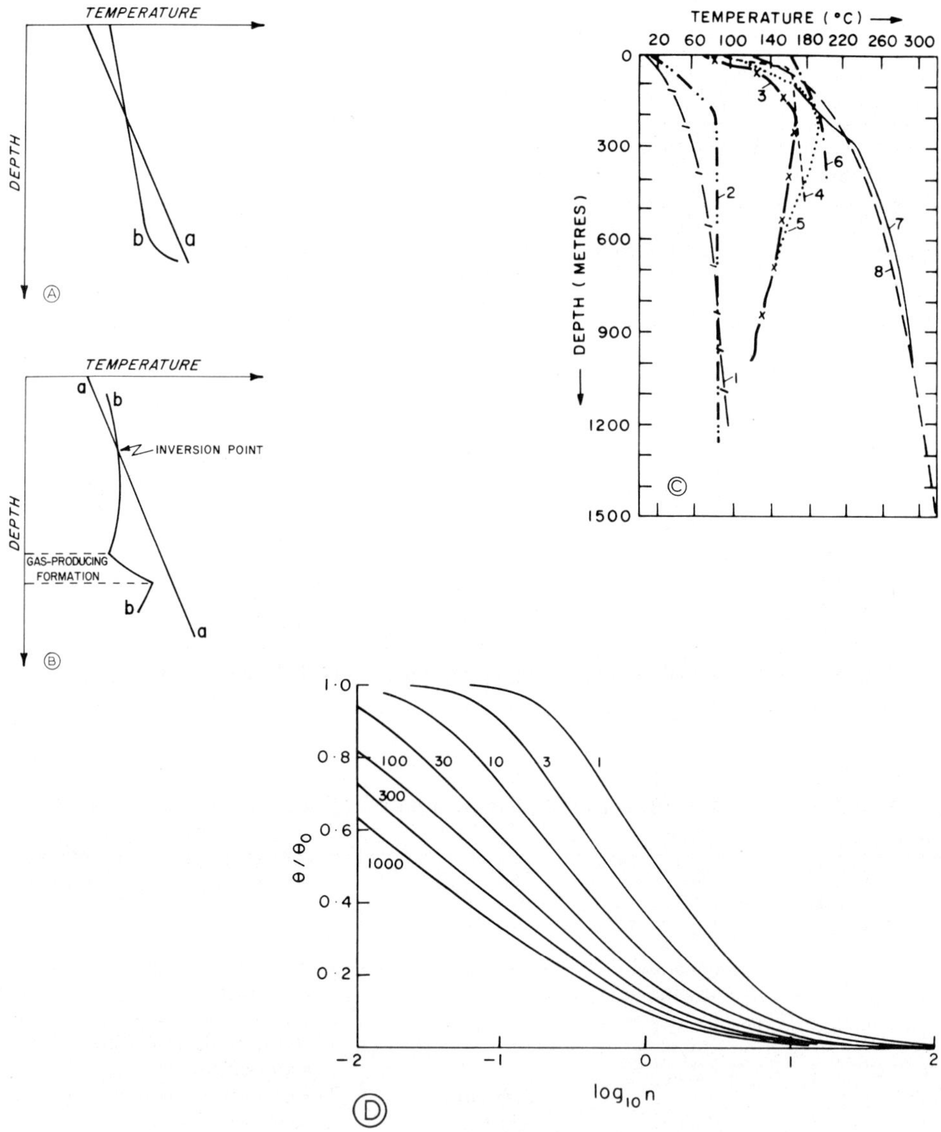

Fig. 6-7. A. Schematic representation of the rock temperatures before drilling (a), and immediately after drilling (b) under the ideal conditions of uniform thermal conductivity of rocks, normal drill hole conditions and continuous drilling. B. Temperature inversion due to a gas-producing zone: a = original temperatures with depth, and b = temperatures when a gas-producing horizon exists. C. Thermographs of wells at Kamchatka geothermal field (*1* through *7*) and the theoretical curve (*8*) for changes in boiling point with increasing normal hydrostatic pressure (from Vakin et al., 1970). D. Recovery in temperature of a borehole whose temperature has been changed by θ_0 for time t_0 and then left undisturbed. The temperature θ at a time nt_0 after the beginning of the undisturbed period is plotted against $\log_{10} n$. The numerals on the curves are ($\chi t_0/a^2$), where χ is diffusivity and a is the radius of the hole.

TABLE 6-3

Classification of thermometers for geothermal thermometry (modified from Ushijima et al., 1976)

Signal transmission and principle of transducer	Thermometer	Example	Features
Self-recording, expansion	liquid in tube	mercury-maximum	(1) cheap, convenient, and simple but fragile (2) long time constant (3) automatic recording and remote sensing are not applicable
	bimetal	geothermograph	(1) an expensive cable is not necessary (2) heat-resistance and waterproofing are very good (3) movement of recording sheet (smoked glass plate) is uncertain (4) continuous measurement is impossible
	Bourdon type	Amerada type	(1) driving of recording chart is done by clock (2) has the same disadvantages as those of the geothermograph
Transmission, electric	thermoelectric	thermocouples	(1) expensive long compensating lead is necessary (2) terrestrial potential field may superpose on emf of thermocouples
	electrical resistance	thermistor	(1) expensive electric cable is necessary (2) reproducibility is poor due to the time deterioration of the semiconductor
		platinum resistor	(3) remote sensing and automatic recording are easy (4) high accuracy is attainable (5) time constant is comparatively small

ward transmission in accordance with the saturation conditions, whereas dry steam and water below 100°C temperature reach the wellhead without any phase conversion. Formation permeability and the physical restrictions at the wellhead control the flow rate. When a well passes through zones producing fluids at the same temperature, the enthalpy of the fluid produced is found to be fairly constant over a wide range of flow rates.

A well flow pressure maximum is reached when the well is choked to limit by a complete closure of the well. The rate of closure affects the maximum discharge pressure. The lower limit of the flow pressure at which a well continues to be operative depends upon the reservoir permeability and the friction encountered in the surface pipework. When the fluid produced by a well is drawn from two or more horizons at different levels, changes in the wellhead throttle influence the proportion of the fluids being drawn from each level.

Under the circumstances of two-phase flow, small and rapid pulsations are most commonly observed. These pulsations are caused by the continuous boiling which takes place as the fluid rises to zones of lowering pressures in the well. Another characteristic of a two-phase flow is the difference in the velocities with which the fluid moves in the pipe section. Water flows with relatively slower velocities along the walls of the pipe while steam and water move much faster in the central part of the pipe cross-section.

Temperature measurements

Thermometers used for measuring temperatures in boreholes can basically be divided into two categories: (1) self-recording expansion type, and (2) electric signal transmission type. The thermometers belonging to the first category either operate on the principle of change in volume of a liquid or on the curving of a bi-metallic sensor with changes in temperature. These thermometers are simple and convenient in operation and do not require an expensive heat-resistant cable. However, they require repetition of the manual operation of lowering and raising in a borehole to investigate temperature changes with time. The thermometers belonging to the second category make use of the thermoelectric effect and change in electrical resistance of metals with temperature. The signal is mostly transmitted through an electric cable, and can be recorded continuously by an analog or digital recorder. It is possible to use radio signals and avoid the use of the cable for transmitting the signal. The salient features of the thermometers used in the geothermal industry are summarized in Table 6-3 (Ushijima et al., 1976).

Pressure measurements

Pressure gauges equipped with Bourdon-tube-type pressure elements are mostly used for measuring geothermal fluid pressure in boreholes. Fig. 6-8 shows schematically a multiple-coil Bourdon tube pressure gauge (the Kuster KPG Subsurface Recording Gauge). The bellow-Bourdon tube fluid-filled system senses the pressure from the well and a change in pressure rotates the Bourdon tube. The stylus shaft transmits this rotation to the stylus assembly. A spring-driven clock rotates the chart carrier loaded with chart paper at a known rate past the stylus. A recording of pressure deflection with time is

Fig. 6-8. Multiple coil Bourdon-tube pressure gauge.

made by the stylus on the chart with no overlapping. The stylus lightly burnishes the grey- or black-coated chart, producing the record. The axial distance on the chart indicates time from which the depth is inferred, and the offsets are governed by the ambient pressures. A calibration table, provided with the gauge, is used in reading pressures off the record.

In the wells characterized by single-phase flow only, the downhole pressures can be deduced from the knowledge of wellhead pressure and the downhole temperature readings (to provide density values). When temperatures have been proved to stabilize with repeated measurements, deduction of downhole pressure from wellhead pressure is quite accurate.

Flow measurements

During the process of drilling a well, a log of the losses of the drilling fluid and of temperature is normally made. The logs are useful in estimating the approximate output of a well, as well as in deciding a suitable method for the measurement of flow. Observations of the variation of wellhead pressure with time provide useful indications of any changes in quality and quantity of flow. Usually the flow is very erratic within the first few hours to a few days of completion of a well.

For a better understanding of the flow characteristics of a well, it is desirable to measure the rate of flow of total mass, the temperature and chemical constituent of single phases. These measurements are made at several wellhead pressures. The wellhead pressures are controlled by throttling the flow with valves or orifices with different diameters. Instead of trying to achieve a specific pressure for flow measurement to be made, it is more efficient to use a convenient choke setting and obtain a graph between resulting pressures and flows. In the event that the well exhibits a smooth relationship, flows at desired pressures can be accurately obtained through graphical interpolation. It is advisable to repeat flow measurements for both increasing and decreasing stages of throttling. However, when changes in output with time are being investigated, it is best to repeat the same process.

Measurement through an orifice. Single-phase flows under pressure are suitably measured by observing the pressure difference across an orifice in a straight run portion of the flow pipe. The pressure difference is measured using a manometer. Standard engineering textbooks carry the necessary information in regard to orifice geometry, flow calculations, possible errors and their mitigation. Since flow measurements using an orifice require single-phase flow, saturated water must either be cooled or pressurized to avoid boiling at the orifice.

Measurements with a calorimeter. Calorimeters are used to measure the volume and heat produced during a certain time interval. The flow is passed through a tank partly filled with water whose thermal capacity and weight are known before the beginning of the experiment. With the measurement of the increase of temperature and volume of water during a known time duration, the heat and flow rates are calculated. Such calorimeters are particularly suitable for small flow measurements. In the case of large flows, a small (say, 1 cm) diameter sampling tube can be inserted to collect the fluid to be measured by the calorimeter. The tube is made to traverse the entire radius of the well. The timing for sampling at each radius should be proportional to the area of flow it represents. This method is inexpensive, but the results obtained are not very reliable.

Cyclone separator vessel. Large steel cyclone separator vessels, like the one shown in Fig. 6-9 (from James, 1976b), are often used to measure the mass flow and enthalpy of a two-phase discharge. The steam and water phases are separated in a vessel with the help of horizontal pipes. Throttle valves are used to control the flow and orifice meters are used to measure the rate of discharge. The method is quite simple and straightforward as long as one phase only passes through each orifice, but the presence of a second phase could cause serious errors. The presence of steam in the pipeline containing the orifice meter being used to measure the flow of saturated hot water is detected by a significant rise in differential pressure across the

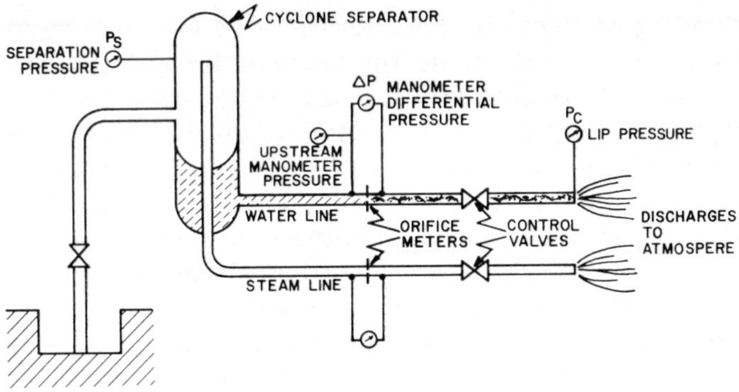

Fig. 6-9. An arrangement for measuring separated water and steam from a geothermal well (from James, 1976b).

orifice. James (1976b) has given empirical relations to estimate the percentage of steam quantities from the percentage increase in manometer differential pressure.

Pressure-transient analysis

The effect of precipitation of solids (and consequent blocking of the drainage in the borehole face) into the well as well as the periphery of the entire geothermal reservoir is important. The precipitation occurs when the cold recharge fluids contact the hot geothermal fluids. This results in decrease in production with time. Steam wells in the Larderello geothermal field in Italy are known to have an active life of about twelve years. The pressure-transient analysis is helpful in determining the reasons for the decline in production often observed in oil, gas and geothermal fields. In the simplest pressure-transient analyses, the static bottom hole pressure is measured at increasing intervals of time after closing the well. The lower the permeability, the longer it takes for the pressures in the well to stabilize. This information is used in determining formation permeability. Fig. 6-10A represents a Miller-Dyes-Hutchinson graph (after Ramey et al., 1976) for a well in the center of a reservoir whose boundaries could be approximated by a square. Further, a full recharge at the boundaries is hypothesized. In the graph, dimensionless, static build-up pressures are plotted against dimensionless shut-in time with the following definitions:

$$P_{DS} = \frac{kd\,(p_i - p_{ws})}{0.4568\, V_{sc}\, qB\mu}, \text{ for liquid flow} \qquad (6\text{-}1)$$

$$P_{DS} = \frac{Mkd\,(p_i^2 - p_{ws}^2)}{0.2789\, qZ\theta}, \text{ for steam or gas flow} \qquad (6\text{-}2)$$

$$t_{DA} = \frac{0.3604\, kt}{\phi \mu C_\theta A} = t_D (r_w^2/A) \tag{6-3}$$

The relationship between slope b and effective permeability k is:

$$k = \frac{0.5258\, V_{sc} q B \mu}{bd}\text{, for liquids, and} \tag{6-4}$$

$$k = 0.3210 \frac{q \mu Z \theta}{bd}\text{, for steam and gases.} \tag{6-5}$$

where:

P_{DS} = dimensionless static pressure
k = effective permeability to the flowing phase, darcies
d = net formation thickness, meters
p_i = initial pressure, kg/cm^2
p_{ws} = bottom hole pressure (static), kg/cm^2
p_{wf} = bottom hole pressure (flowing), kg/cm^2
p_{ts} = surface static pressure, kg/cm^2
V_{sc} = specific volume under standard conditions, cm^3/g
q = production rate, 1000 kg/hr
B = formation volume factor (reservoir volume/standard volume)
μ = viscosity of flowing fluid, 10^{-3} P
M = molecular weight, g/g-mole
Z = real gas law deviation factor
θ = absolute formation temperature, °K
ϕ = porosity in fraction of the bulk volume
C_θ = total system effective isothermal compressibility, cm^2/kg
r_w = well radius, meters
A = drainage area, square meters
b = slope of the semilog graph, (kg/cm^2)/log cycle for liquids and (kg/cm^2)^2/log cycle for gases
t = time, hours
t_D = dimensionless time
t_{DA} = dimensionless time based on drainage area, A.

In Fig. 6-10B a Horner graph for a well in the center of a constant-pressure square, i.e., fully recharged, is shown (from Ramey, 1976). An infinitely long shut-in time (Δt) would make $[(t + \Delta t)/\Delta t]$ approach unity and a dimensionless build-up pressure P_{DS} of zero since the well would return to the initial pressure. The build-up lines appear to move to the right with the increase of producing time prior to shut-in increases.

Horner-type build-up graphs are particularly useful in geothermal well measurements. Investigations of the *skin effect* or *skin factor* are useful in

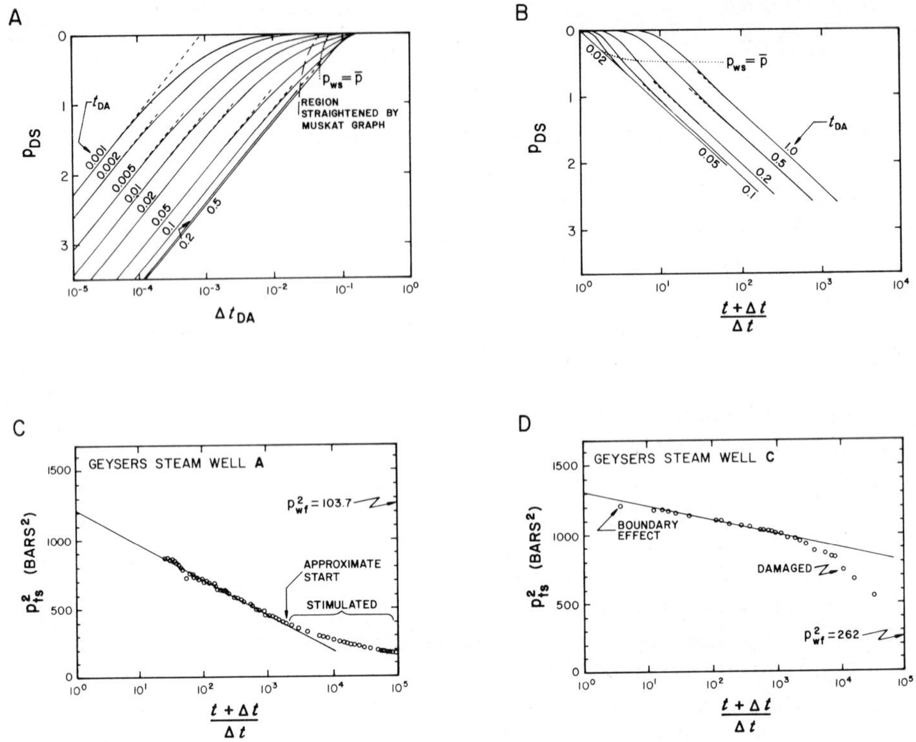

Fig. 6-10. Graphs commonly used in pressure-transient analyses, and their application to detect stimulated and damaged conditions at The Geysers. Miller-Dyes-Hutchinson graph (A), and Horner graph (B) for a well located at the center of a constant pressure square. Detection of stimulated (C) and damaged conditions (D) in steam wells at The Geysers. See text for details and the notations used.

determining whether a well is damaged, stimulated or is normal. On the Horner graph, pressure build-up with time exhibit semilog straight lines. A deviation is an indication of stimulated or damaged situation. Fig. 6-10C and D exhibit pressure build-up graphs for two steam wells at The Geysers (adapted from Ramey, 1976) exhibiting stimulated and damaged conditions respectively.

In conclusion it may be mentioned that with passage of time, more realistic models for pressure-transient analyses are being made and they are very useful in estimating well conditions.

Reservoir modelling

The geothermal reservoirs are usually complex hydrothermal systems involving single- or multiphase fluid flow. Mathematical and/or laboratory simulation of these reservoirs is helpful in estimating the recoverable energy, in determining the optimum management techniques and improving the understanding of the reservoir geometry, and in estimating boundary conditions and rock properties. As the production progresses, newer data permit refinement of initial crude models and a more purposeful and increasingly more accurate prediction of the reservoir performance.

Mathematical modelling

Simple mathematical modelling basically consists of the formulation of continuity equations of mass, momentum and energy for each phase and the reduction of this system of equations to two non-linear partial differential equations. The final simplified equations, whose solution is sought, have two unknown dependent variables. These variables can be chosen from fluid pressure, temperature, enthalpy, water saturation, etc. As an example, let us discuss a case where fluid pressure and enthalpy have been chosen as the two unknown variables (Faust and Mercer, 1976). Consideration of the thermodynamic state of water under simplifying assumptions allows total density, saturated steam density, saturated water density, saturated steam enthalpy, saturated water enthalpy, water saturation and temperature to be expressed as functions of fluid pressure and enthalpy (Fig.6-11). Next, the partial differential equations are transformed to approximate integral equations and solved simultaneously using a finite element technique. The dependent variables, pressure and enthalpy, are approximated using piecewise polynomials.

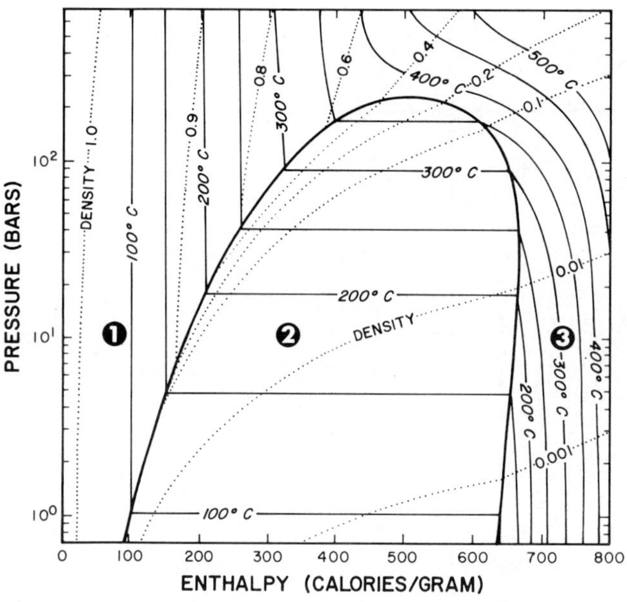

Fig. 6-11. Pressure-enthalpy diagram for pure water and vapor showing the three thermodynamic regions below the critical point (apex of parabola). *1* = compressed water, *2* = two-phase steam-water, and *3* = superheated steam (from White et al., 1971).

Steady state, vertical, one-dimensional flow. Donaldson (1968) has considered one-dimensional vertical flow of water and heat in a porous medium invclving compressed water with the boundary conditions: at the top of the system ($Z = 0$), the temperature is θ_1 and the pressure is p_1; at the base ($Z = Z_2$), the temperature is θ_2 and a mass flow rate per unit area is specified as M. The analytical solution for temperature assuming constant permeability, k, thermal conductivity, K, and heat capacity of water Q_w is:

$$\theta = \theta_1 + (\theta_2 - \theta_1) \frac{1 - \exp\left(-\dfrac{MQ_wZ}{K}\right)}{1 - \exp\left(-\dfrac{MQ_wZ_2}{K}\right)} \tag{6-6}$$

The fraction MQ_w/K indicates the relative importance of convection in comparison with conduction. With convection becoming more dominant, the value of the fraction increases. In Fig.6-12 results obtained for different values of MQ_w/K by Donaldson (1968) are presented. Additional data used in arriving at these results are: a thickness of 3 km; $\theta_1 = 116°C$; $\theta_2 = 224°C$; $p_1 = 3.0 \times 10^7$ dynes/cm^2; $M = 3.2 \times 10^7$ ergs/g °C; and $K = 3.2 \times 10^5$ erg/cm s °C.

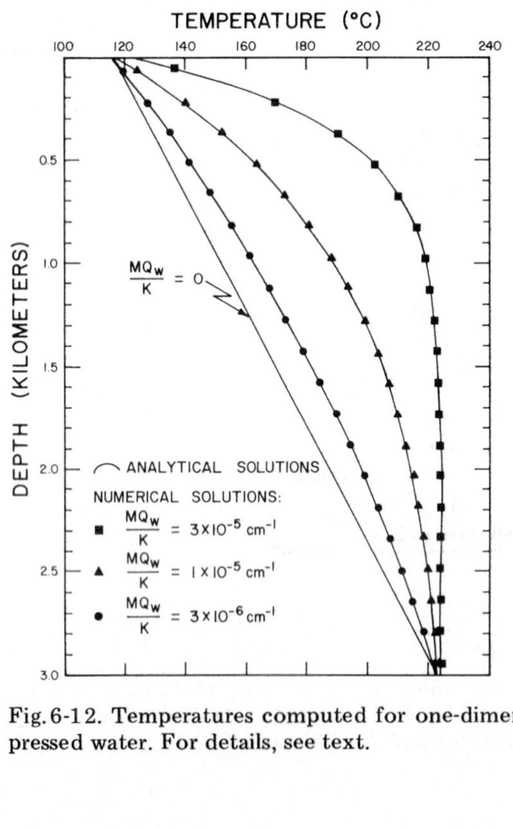

Fig. 6-12. Temperatures computed for one-dimensional steady state vertical flow of compressed water. For details, see text.

Multiphase horizontal flow. Faust and Mercer (1976) have mathematically investigated the behavior of an initially hot water geothermal reservoir which developes a two-phase (steam and water) zone under the influence of production. Results obtained by them are shown in Fig. 6-13. In Fig. 6-13A pressure values at various times are plotted against distance from the center of the reservoir and the times specified correspond to the duration of production. It is assumed that the production well bottoms in the center of the reservoir. During the initial stages of production, the pressure drops rapidly in the entire reservoir. After a month, the fluid splits into hot water and steam at the well bottom and consequently the pressure is maintained there while it drops in the rest of the reservoir. After three years of production, the pressures in the entire reservoir have lowered to a value which is just above the pressure where two phases could exist. Further production reduces water saturation in the production zone. Fig. 6-13B indicates water saturation after five years of exploitation at various distances from the production well.

In addition to investigating multiphase flow, convection pattern, heat and mass transfer and other problems concerning the dynamics of geothermal

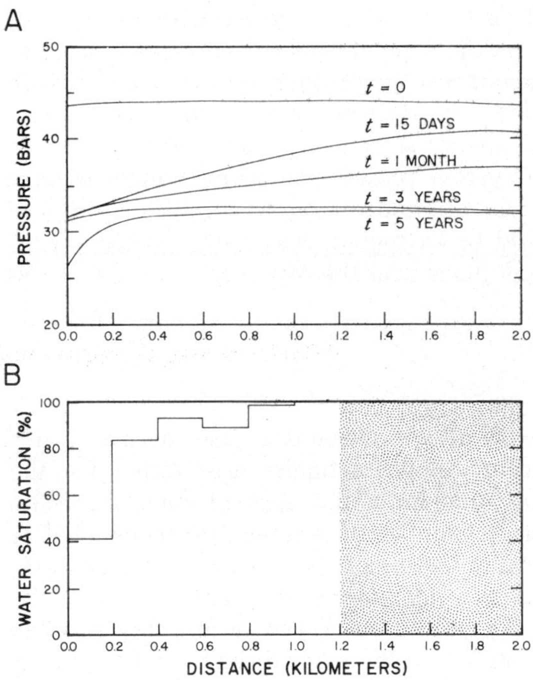

Fig. 6-13. Behavior of an initially hot water geothermal reservoir which develops a two-phase zone under the influence of production (from Faust and Mercer, 1976). A. Computed pressures versus distance from the center of the reservoir (left boundary of the model) at various times. B. Computed water saturation versus distance from the center of the reservoir after five years of exploitation. Shading indicates water saturation.

reservoirs, the effect of production rate on the ultimate heat produced and on the behavior of the ratio of steam to water production have been mathematically investigated. Robinson and Morse (1976) report that the total amount of heat that may be extracted from a geothermal reservoir is limited, and to obtain the maximum amount of heat, the reservoir pressure should be reduced to a point where the superpressured water will flash to steam within the pores of the rock.

Laboratory modelling

Many parameters, important in estimating the production potential of geothermal fields, can be investigated through laboratory modelling. Results of a recent experimental study are briefly mentioned here.

In many known geothermal areas porosity and permeability are too low to sustain adequate steam production rates. Also, as mentioned in Chapter 1, a large portion of the resource base consists of hot rock formations with no water. The extent to which geothermal energy stored in a hot rock can be extracted under varying conditions is not known. To investigate the production potential of hot dry rock under various circumstances, Hunsbedt et al. (1976) carried out experiments on a laboratory-size model simulating a fractured rock chimney. The model produces steam, either with or without simulated geofluid recharge, from initial temperature and pressures of up to 260°C and 55 bars. Transient behavior resulting from steam production with two rock loads having 44% and 35% porosity was investigated. These experiments demonstrated that thermal energy stored in a fractured rock can be extracted effectively by reducing system pressure to allow boiling to take place in the rock formation. Under optimum operating conditions, 90% of the geothermal energy stored could be extracted. Maximum energy extraction was achieved when boiling took place near the top of rock mass and was controlled by recharge.

PRODUCTION TECHNOLOGY

Production technology deals with the movement of geothermal fluids from their sources to the sites where they are finally used either for the generation of electric power or for other industrial, agricultural or domestic purposes. Between these two ends, a number of practical problems dealing with the choice of proper equipment, designing of the transmission pipework, and taking preventive measures against corrosion and scaling need to be tackled. Additionally, practices and methods employed to stimulate production are a closely related topic, which is discussed briefly in Chapter 9.

As mentioned earlier, the characteristics of the fluids produced vary considerably from one geothermal field to another. At places, only steam or high-temperature water with low impurities is produced, while at others a mixture of steam and water with a high percentage of incondensible gases like H_2S and CO_2 is produced. Sometimes water dissolved with a variety of other chemicals, whose recovery may be profitable, is produced. During the initial stages of production, soon after drilling has been completed, rock fragments and debris are ejected. Gradually, the well blows itself clear of these drilling remains, but small particles from the formations may continue to be entrained by the fluid produced. The chemical properties of the fluid produced, the effect of changing wellhead pressure on the rate of production, the ratio of steam to water, as well as the maximum wellhead pressure in a closed well, are the characteristics which need to be known before designing an efficient collection and transmission system. Usually a geothermal field has a number of producing wells, and the fluids produced by wells at similar pressures may be mixed, forming separate transmission lines for each pressure group. The fluids may be transmitted as hot water, steam, or a mixture of the two when separation is not done at the wellhead.

Wellhead equipment

An arrangement for typical wellhead equipment is shown schematically in Fig. 6-14. Closed wells have much higher wellhead pressures compared to those prevailing under normal production conditions. To relieve very high pressures, a bleed valve is usually installed below the wellhead master valve. The shut-off valve sometimes experiences abnormally high forces due to the impact of geothermal fluid moving at high velocities at the time of closure. Damage is prevented by using valves with large seating surfaces. It is advisable to operate the master valve only when the flow is completely cut off and regulate the flow with other downstream valves. The master valve should be used as the last resort when the other valves malfunction. When the fluid produced is a mixture of steam and water, a cyclone separator, like the one shown in Fig. 6-9, is commonly used to separate the two phases. The cyclone

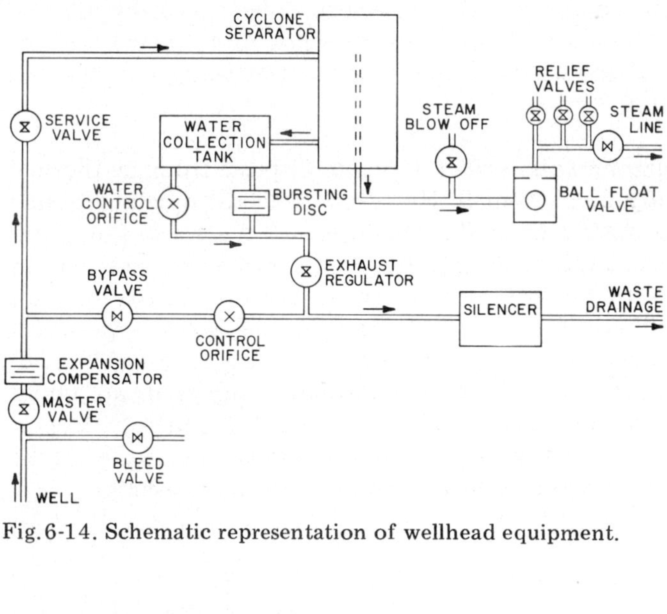

Fig. 6-14. Schematic representation of wellhead equipment.

separator consists of a vertical cylinderical vessel in which the mixture of steam and water enters tangentially through a spiral inlet pipe. The water swirls around the wall and moves downward under gravitational force, while steam collects at the top and leaves through a pipe emerging from the bottom. The water level is controlled by an orifice which is helpful in preventing water from entering the steam line. In the scheme shown in Fig. 6-14, the separated water is discharged to waste. However, it could be used for the purposes of district heating, greenhouses, etc., or steam at low pressures could be obtained from it by flashing and passing it through a second-stage cyclone separator. The ball float valve prevents possible carry-over water from entering the steam main. When the water in the valve reaches a critical level, the ball rises and closes the steam flow. Consequently, the pressure rises and the bursting disc ruptures, causing the excess of water to flow out. It is useful to provide a bypass line capable of discharging the total flow from the well. This provides a means of isolating and inspecting the wellhead equipment without closing the well, as well as stabilizing the flow before diverting it to the separator.

Noise reduction. When waste fluids are allowed to escape to the atmosphere through an open-ended pipe, the noise level could be as high as 140 decibels (dB). Besides being a nuisance, this noise can affect the hearing of people at work. With the use of an efficient silencer the noise level can be reduced to 100 dB. It can be reduced most effectively by discharging the waste under water. However, for this process, the proximity of a lake or some other suitable body of water is necessary. At low noise levels, horizontal silencers, in which the pipe diameter increases gradually, are useful. Twin-cyclone silencers are very efficient at high noise levels. In these silencers, the entering fluid mixture is bifurcated by a flow splitter and made to flow in clockwise and anti-clockwise rotational swirls. Friction and turbulence destroy most of the kinetic energy of the fluid. The steam expands and emerges at low velocity from the top of the silencer, whereas swirls of water impinge and discharge from the bottom.

Waste disposal. The environmental effect of waste disposal from geothermal fields is dealt with briefly in Chapter 9. However, it needs to be mentioned here that the problems arising from the chemistry of the waste hot water should be given due consideration during the planning of the project. Wherever acceptable, disposal of the hot water in rivers, lakes of the sea provides the least problem. Another alternative is to reinject the waste water back into the formations. A suitable drainage system connecting each wellhead to the main drainage channel becomes necessary where many wells are under operation. The drainage channels are usually constructed of concrete. During drainage, the hot water gradually deposits considerable quantities of its mineral contents. Open channels make the cleaning operation easier.

Transmission of hot water

One of the most important considerations in hot water transmission is to maintain adequate pressure so that the water does not boil and formation of steam is suppressed. The consequence of lowering and subsequently rising of the pressure could be disastrous. When the pressure is lowered, pockets of steam as well as air bubbles are formed. These steam pockets and bubbles collapse when the pressure is increased again. This causes a *water hammer*, which could burst the pipe. An adequate margin of pressure should be maintained by increasing the pressure through pumping, or by adding cold water, or through a combination of both. The pressure also drops gradually due to friction in the pipe. Topographic elevation changes of the pipeline also alter the pressure.

Surging is another important aspect of hot water transmission which deserves considerable attention. Surging occurs when the velocity of the moving water changes as a result of operating the control valve at the delivery end of the transmission pipeline. Opening or closing of this valve causes pressure oscillations. A rapid closure would suddenly decelerate the mass movement, consequently causing an increase in pressure. Similarly, a rapid opening would cause a pressure decline which would result in the formation of steam, which would later on collapse. Hence, it is important to investigate the *surge effects* systematically and to change the delivery end opening so that the resulting pressure variations are within the tolerance of the pipeline network.

Transmission of steam

One of the chief considerations in designing a pipeline for the transmission of steam is the choosing of an appropriate diameter for the pipe so that there is not an excessive drop of pressure between the wellhead and the delivery end. In a typical geothermal field, the steam produced at several wellheads is transmitted through thinner pipes to the main transmission line. An optimum pressure, at which transmission should be performed, would mostly depend on the output characteristics of a group of wells feeding one particular transmission line. The density of steam effects the pressure drop during transmission. For the same pressure drop, low-density steam can flow at higher velocity than a high-density steam under other similar conditions. However, very high velocities can erode and damage the pipe. Mostly the velocities are kept in the 20- to 50-m/s range for transmission of saturated steam and are not permitted to exceed 60 m/s, even in large-diameter pipes. Additionally, pressure drops are due to frictional resistence of flow and due to sudden enlargements or contractions of the diameter of pipes at the joints. For a given pressure and length of the pipe, the pressure drop is inversely proportional to the internal radius of the pipe and directly proportional to the square of the velocity. Uchiyama and Matsuura (1970) used the following empirical relation in deciding the inside pipe radius for steam transmission for the Matsukawa geothermal power plant in Japan:

$$r = \frac{1}{2}\left(\frac{0.785\ wv}{3600\ U}\right)^{1/2} \tag{6-7}$$

where r = inside radius of the pipe, w = amount of steam being discharged, v = specific volume of 30% moist steam, and U = velocity.

Heat losses can be minimized to some extent by providing an insulated covering on the pipeline. Unless the steam is superheated, some of it condenses during its passage and it is useful to remove the water through *catch pots* installed at intervals. This also helps reduce the salt content which, when present in excess, is damaging for the turbine blades. As a precautionary measure against the malfunctioning of the equipment intended for preventing water from entering the steam mains, float-switches can be installed in the *catch pots* to detect the presence of excessive water and issue a warning signal.

Usually the wellhead valves are kept fully open and the pressure is controlled from the delivery end of the pipeline. Thus the wellhead pressure would be higher than the delivery end pressure by an amount equal to the pressure loss during transit of steam and flow at each well would be automatically adjusted. Under extraordinary conditions, the wellhead pressure may be very high, necessitating its control through a throttle.

To ascertain that enough steam is available all the time in a central steam pipeline, usually more than the minimum number of wells required are connected. The excess steam is discharged through a vent valve which also provides the necessary buffer to accommodate the fluctuations in steam requirements. A large reduction in demand causes excessive pressure increase, which is relieved through the safety valves provided at the wellhead.

Additional engineering details for the design of steam pipelines, as used in the geothermal fields of Japan and Italy, are given by Uchiyama and Matsuura (1970) and Pollastri (1970), respectively.

Transmission of a steam-water mixture

There are many advantages in transmitting a mixture of steam and water from wellhead to the powerhouse in comparison with separate single-phase transmission. This scheme is particularly very useful when the hot water produced is intended to be utilized. The main advantages are due to the elimination of much equipment used for separating the phases at the wellhead, no need to suppress boiling in the pipeline, and generation of more power from the same amount of fluid produced.

As outlined by Smith (1973), depending upon pressure, temperature and ratio of water and steam, the flows can broadly be classified as:

Bubble flow: bubbles of steam move along the upper part of the pipe at approximately the liquid velocity.

Plug flow: alternate plugs of steam and water move along the upper part of the pipe above the water.

Stratified flow: water flows in the bottom of the pipe with the steam above, over a smooth interface.

Wave flow: due to higher velocity, the interface has become disturbed by waves travelling in the direction of flow.

Slug flow: a slug of water is picked up periodically by the more rapidly moving steam and continues at much greater velocity than the mean velocity of the water.

Annular flow: the water forms in a film around the inside of the pipe and the steam flows at high velocity in a central core.

Spray flow: nearly all of the water is entrained in the steam.

It had been the general belief that concurrent transmission of steam and hot water, even in steady state, would cause excessive vibration, water hammer or surging because of instability. However, a series of tests conducted in New Zealand (James et al., 1970) and Japan (Takahashi et al., 1970) provided favorable results and no appreciable vibration, water hammer or surging in the steady state transmission was noted. The pressure drop of the steam-water mixture was moderate and acceptable from the point of view of practicality.

More encouraging results have been recently obtained from experiments conducted on two-phase transmission. Soda et al. (1976) conducted experiments at Hatchobaru, Japan with a 30-cm-diameter, 200-m-long pipe arranged sloping downwards at an angle of 7° and provided with a shut-off valve at the delivery end. They report the pipeline to be sufficiently stable without any sounds or vibrations, even when the flow of mixture was shut off quickly. The stability was confirmed with visual as well as oscillographic testing of the pressure fluctuations. The sudden rise of the pressure at a point just before the shut-off valve was found to be lower than that predicted from theoretical considerations. At ordinary velocities, the flow was found to be of semi-annular mist type, whereas for lower velocities it was found to be stratified. The inside surface of the pipe was inspected after 74 days of experimentation and was found to have very little deposition and the surface was found to be free from erosion and corrosion at the moderate flow rates of the mixture.

These results indicate that in the near future, wherever possible, two-phase flow of geothermal fluids could be favored in designing the transmission lines.

Scale deposition and corrosion control

Geothermal fluids are generally contaminated with chemical impurities of underground origin and may further be contaminated with impurities from the atmosphere. These cause scale deposition as well as erosion of the various equipment, piping and disposal channel surfaces that come in contact with the geothermal fluids. A knowledge of the amount of the chemical impurities present in the geothermal fluids and their corrosional and depositional effects on various surfaces under varying conditions of temperature, pressure and fluid movement velocity are helpful in preventing the damage caused by deposition and erosion. Under extreme conditions, continuous scaling could choke the pipelines and continuous corrosion could make sections of pipelines and other equipment weak and incapable of handling the pressures they were originally designed to withstand.

Scale deposition prevention research

Up to the present, geothermal brines are mostly discarded because of their scaling effect, in spite of their useful heat energy content. Experiments are now being conducted to determine the relationship between the brine chemistry, process conditions and scale deposition. Wahl and Yen (1976) have reported the results of experiments conducted using brine from the East Mesa geothermal field, California. They found that deposition of calcite occurred at a rate which depended on the flashing of the brine and separation of the vapors but was independent of the heat flux. An examination of the chemical and the physical nature of the deposition indicated that it was quite sensitive to process conditions. The effect of electrical potential on scale formation has been investigated at the Lawrence Livermore Laboratory, California (Schock and Duba, 1975). The scale formed at the Salton Sea geothermal field, California was found to consist mainly of silicon, iron, copper, silver and aluminium. Significantly more scale was found to deposit on negative electrodes as compared to the positive or the neutral electrodes. The scale formed on the negative electrode contained up to 1000 times more lead than the other electrodes. The amount of scale formed on the positive electrode appeared to be the same as on the negative electrode, however, the scale on the former was much richer in iron and zinc. The results of these experiments indicate that a system could be designed where, through the control of acidity and/or oxidation, the scale-forming constituents could either be kept suspended in the fluid or collected in a separator. In another investigation carried out at the Lawrence Livermore Laboratory (Jackson and Hill, 1976) on preventing scales formed by sulphides, it has been found that the problem could be solved through partial oxidation or combined oxidation-acidification processes.

In addition to the experiments conducted on chemical and electrical modification of the fluid to prevent scale formation in flowing geothermal brines, the other preventive procedures researched include fluid pretreatment to remove scale-forming constituents and hardware surface treatment to prevent adhesion and chemical or mechanical removal. It is hoped that in the near future these research efforts will provide efficient and inexpensive scale-preventing methods and that it will be possible to use geothermal brines for energy production economically.

Corrosion control

Marshall and Braithwaite (1973) have given a good account of the corrosion problem in the geothermal energy production industry and its control. Here we shall summarize their findings and the results of some other recent experiments carried out to control corrosion. Tables 6-4 through 6-6 are useful in selecting suitable material for constructing equipment and specific items to be used for handling geothermal fluids.

Surface corrosion. Geothermal fluids containing free sulphuric, hydrochloric or hydrofluoric acid cause extensive surface corrosion, thereby prohibiting their practical use. However, geothermal fluids discovered and used so far do not contain these acids in proportions which may prohibit their use, and provided the fluids are not permitted to be contaminated by atmospheric oxygen, their corrosional effect on common structural material is not high. Aeration of geothermal fluids causes depolarization and drastically increases their corrosional effect. Austenitic stainless steel, titanium, and chromium (plating) are among a few exceptional metallic alloys which can be used in engineering construction. Adequate protection against corrosion by aerated geothermal fluids can also be provided by covering a steel surface with an epoxy coating.

Erosion-corrosion. The combined action of erosion and corrosion on metals is significant, particularly in items like the blades of the turbine where wet steam at high velocity could be very damaging. Experiments show that a 13% chromium-stainless steel blading alloy provides adequate erosion-corrosion resistance. Corrosion-resistant alloys, such as austenitic stainless steels, provide good resistance against erosion-corrosion.

Stress corrosion. The presence of chloride solutions at high temperatures and the presence of tensile stresses cause stress corrosion. There does not seem to be a lower stress limit below which stress corrosion cracking would not occur. It appears that austenitic stainless steels are not susceptible to stress corrosion in air-free geothermal fluids. However, the presence of dissolved oxygen in geothermal fluids would cause stress corrosion of austenitic stainless steel.

Sulphide stress cracking. Experimental investigations and observations in the field have shown that even high-strength steels are susceptible to cracking when simultaneously stressed and exposed to aqueous solutions of H_2S. Since the presence of H_2S is commonly observed in geothermal fluids, sulphide stress cracking is a factor of major importance in the design and selection of material for equipment. Medium- and high-strength carbon and alloy steels are not very suitable except when the environment is known to be free from a high concentration of H_2S.

Hydrogen infusion. It has been discovered that infusion of hydrogen into steel occurs when geothermal fluids containing aqueous solutions of H_2S corrode the steel surfaces. Under favorable circumstances, hydrogen fusion

TABLE 6-4

Summary of corrosion test results in non-aerated steam (150 days) (after Tolivia et al., 1976)

Materials	Corrosion rate (mm/yr)	Corrosion rate law*	Characteristics**	Maximum depth of pits or cavities (mm/yr)	Stress corrosion
12Cr steel	0.010 (0.004 ~ 0.013)	L	SP-6	<0.024	not sensitive
12Cr-1Mo-1W steel	0.006 (0.004 ~ 0.010)	P	SP-6	<0.024	not sensitive
12Cr-0.2Al steel	0.021 (0.0005 ~ 0.023)	RP	GC-2	general corrosion	not sensitive
15Cr-1.7Mo steel	0.0045 (0.0015 ~ 0.0075)	L	SP-1	0.05	not sensitive
1Cr-1Mo-1/4V steel	0.044 (0.022 ~ 0.059)	I	GC-2	general corrosion	sensitive
3.5Ni-1 3/4Cr-0.5Mo-0.1V steel	0.020 (0.011 ~ 0.038)	L	SP-2	0.12	sensitive
1.5Cr-1Al-1/4Mo steel (nitrided)	weight gain	—	nitrided layer were removed partly	0.97	not tested
Carbon steel ASTM-A285	0.038 (0.038 ~ 0.13)	P	GC-2	general corrosion	not tested
Aluminum	weight gain	—	GC-1	general corrosion	not sensitive

*L = linear, P = parabolic, RP = reverse parabolic, I = irregular, SA = slightly accelerate.

**SP-1: slight scattered pitting corrosion in wide range
SP-2: scattered pitting corrosion in wide range
SP-3: scattered pitting corrosion in remarkably wide range
SP-4: remarkable scattered pitting corrosion in wide range
SP-5: remarkable densely scattered pitting corrosion in wide range
SP-6: scattered pitting corrosion in local range
SP-7: densely scattered pitting corrosion in local range
GC-1: slight general corrosion with uneven surface
GC-2: general corrosion with uneven surface
GC-3: remarkable general corrosion with uneven surface
GC-4: general corrosion with wavelike uneven surface

TABLE 6-5

Summary of corrosion test results in aerated steam (150 days) (after Tolivia et al., 1976)

Materials	Corrosion rate (mm/yr)	Ratio to corrosion rate in non-aerated steam	Corrosion rate law*	Characteristics*	Maximum depth of pits or cavities (mm/yr)	Stress corrosion
12Cr steel	0.09 (0.085 ~ 0.14)	9.0	P	SP-5 microscopic cracks	1.7	sensitive
12Cr-1Mo-1W steel	0.09 (0.04 ~ 0.23)	15	P	SP-4 microscopic cracks	1.6	sensitive
12Cr-0.2Al steel	0.11 (0.06 ~ 0.16)	5.2	P	GC-3 microscopic deep pitting	general corrosion	not sensitive
15Cr-1.7Mo steel	0.019 (0.005 ~ 0.023)	4.2	L	SP-6	1.2	not sensitive
1Cr-1Mo-1/4V steel	0.16 (0.13 ~ 0.50)	3.6	P	GC-3	general corrosion	sensitive
3.5Ni-1 3/4Cr-0.5Mo-0.1V steel	0.20 (0.12 ~ 0.52)	10	P	SP-2 inter-granular corrosion	0.70	—
1.5Cr-1Al-1/4Mo steel (nitrided)	weight gain	—	—	SP-6	0.85	—
Carbon steel ASTM-A285	0.11 (0.065 ~ 0.44)	2.9	L	GC-3	general corrosion	—
Aluminum	0.065 (0.01 ~ 0.095)	—	SA	SP-3	2.9	—
Copper (deoxidized)	0.57 (0.51 ~ 1.11)	—	P	GC-5	general corrosion	not sensitive

*See footnotes of Table 6-4.

TABLE 6-6

Final evaluation of corrosion resistance in condensate (150 days) (after Tolivia et al., 1976)

Material	Velocity of condensate	Condensate rate law*	Average corrosion rate weight loss (mm/yr)	Maximum local penetration rate (mm/yr)	Corrosion characteristics*	Stress corrosion
Deoxidized copper	high	L	0.64	—	GC-5	—
	low	L	0.24	—	GC-2	not sensitive
Aluminium	high	RP	0.48	3.65	SP-4	—
	low	P	0.01	1.16	SP-2	not sensitive
Naval brass	high	RP	0.22	—	GC-2	—
	low	P	0.07	—	GC-2	not sensitive
18-8 stainless steel	high	—	0.0003	—	no corrosion	—
	low	—	0.0008	—	no corrosion	not sensitive
12Cr steel	high	L	0.090	0.85	SP-6	—
	low	P	0.015	0.97	SP-6	—
Low carbon steel	high	P	0.66	—	GC-5	—
	low	P	0.29	—	GC-4	not sensitive

*See footnotes of Table 6-4.

causes blistering, renders steel brittle, and is found to be associated with sulphide stress cracking and delayed fracture of stressed steels. The hydrogen permeation rate falls off rapidly with time at temperatures in excess of 100°C. It is estimated that after 2 to 30 weeks, the corrosion product developed at the exposed surface prevents permeation of hydrogen. Surface coatings with paint or chemically formed magnetite coating and oxidation are helpful in reducing hydrogen infusion. The decrease in permeation activity with the increase of temperature is probably due to change in the corrosion mechanism at high temperatures.

Infusion-induced delayed fractures. Hydrogen infusion is found to be the cause of delayed fractures observed in high-strength steels subjected to tensile stress. Detailed field experimental data on delayed fracture of bore-casing steels exposed to geothermal fluids for a long time under closely controlled tensile loading have been provided by Marshall and Tombs (1969). In these tests, no delayed fracture occurred at high temperature (>100°C); however, when exposed to cold geothermal steam condensate, both low- and high-strength steels exhibited delayed fracturing. Results of experiments on delayed fracture under controlled stress distribution and stress corrosion suggest the use of steels with strengths in excess of 88,000 pounds per square inch (psi) in geothermal steam media devoid of condensate or water.

Recent field data on corrosion investigations. Tolivia et al. (1976) report on the corrosion resistance of different turbine materials in geothermal steam environment. These investigations were conducted at Cerro Prieto geothermal field in Mexico. Fig. 6-15 schematically explains the experimental setup. Steam was separated from the hot water by a separator and introduced into steam and aerated steam chambers (Fig. 6-15A, B). The separated steam was lead through a cooler to obtain condensate and the condensate was stored in a tank and from there introduced into low- and high-velocity

TABLE 6-7

Corrosion test conditions in separated steam and condensate (Tolivia et al., 1976)

	Velocity (m/s)	Pressure (psi)	Temperature (°C)
Separated steam			
non-aerated	<10	70 ~ 194	152 ~ 194
aerated	<10	atmosphere	100
Condensate			
high velocity	0.5	—	38 ~ 58
low velocity	0.02		

test chambers (Fig. 6-15C). The test conditions and chemical compositions are listed in Tables 6-7 and 6-8, respectively.

General corrosion tests were performed on plate pieces measuring 3 cm × 6 cm × 0.3 cm. At intervals of 30 days exposure, for a total duration of up to 150 days, two test pieces of each material were removed from the test chamber and corrosion rates were measured. Stress corrosion tests were conducted on U-bend test pieces to which yield strength was applied. Visual and microscopic observations were made to check for the presence of cracks. Corrosion-fatigue tests were conducted on shank-type plane-bending specimens. Summary results of these experiments are presented in Tables 6-4 through 6-6.

It has been concluded from these experiments that the carbon-steel used in turbine shells and piping has satisfactory corrosion resistance in non-aerated steam, and its corrosion rate increases in aerated steam and conden-

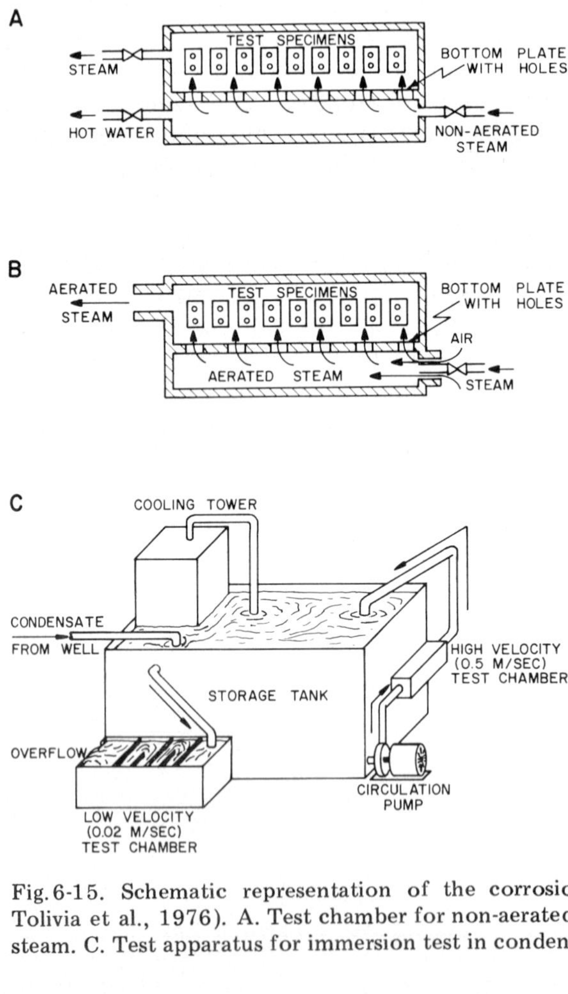

Fig. 6-15. Schematic representation of the corrosion test equipment (modified from Tolivia et al., 1976). A. Test chamber for non-aerated steam. B. Test chamber for aerated steam. C. Test apparatus for immersion test in condensate.

TABLE 6-8

Average chemical compositions of separated steam and condensate (Tolivia et al., 1976)

		Chemical composition (ppm by weight)
Separated steam	CO_2	19,500
	H_2S	2580
	Cl^-	13.3
	SO_4	6.8
	Na	1.29
	K	0.58
	pH	8.35
Condensate	pH	7.1
	Cl^-	42
Conductivity		648

sate. Hence adequate corrosion allowance should be made in design, and coating with epoxy resin would be required in low-temperature conditions. The deterioration of the fatigue endurance limit is more important than the general corrosion rate for the 12Cr steel used for the bucket. For making the rotor, low-alloy steels are almost as good as carbon steel in corrosion rate. Dioxized copper and aluminium are very poor in corrosion resistance in the condensate and as such cannot be used for constructing thin-walled heat exchanger tubes and the use of highly resistant material like titanium is desirable for their efficient functioning.

Corrosional effect on non-metallic material. Not much is yet known about the corrosional effect of geothermal fluids on non-metallic material. However, concrete and grout are widely and successfully used unless atmospheric oxidation of hydrogen sulphide occurs. Results reported by Lorensen et al. (1976) on experiments conducted to investigate the corrosive effect of geothermal brines at 300°C on polymeric and other composite materials, show that polymers containing substantial aromaticity, fluorine, carbonized structure, or aliphatic unsaturation have a high resistance against brines. The presence of water-sensitive groups such as imide, amide, or ester affects even thermally very stable polymers, rendering them unsuitable. In field tests carried out with flowing brine, a fluorocarbon polymer is reported to be very suitable against erosion and scale deposition by Lorensen et al. (1976).

Chapter 7

THE CERRO PRIETO GEOTHERMAL FIELD, MEXICO

INTRODUCTION

The Cerro Prieto Geothermoelectrical Project, Mexico, is one of the very successful ones generating electric power from a geothermal resource. With an installed capacity of 75 MW of electrical power generation, it finds a place among the four largest geothermal power plants in the world. Construction of a similar power station is under way to double the power generation capacity of the field.

The teamwork of the General Managements of Planning and Program, of Construction and of Operations of the Comision Federal de Electricidad, Mexico, are responsible for the development of various phases of the Cerro Prieto Geothermoelectrical Project. The General Management of Planning and Program, through the Department of Geothermal Resources, is in charge of research, development and maintenance of the wells and of the geothermal fields as a whole. The General Management of Operation is in charge of the operation of the generator plant and distribution of electricity in the state of Baja California. The General Management of Construction is responsible for the direction and supervision of the civil engineering works and the assembly of the geothermoelectric plant. Mexican as well as American, British, Belgian, Japanese, Swiss, Italian and French companies have participated in various phases of the development of the Cerro Prieto Geothermoelectrical Project.

The Comision Federal de Electricidad, a government agency, is responsible for power generation in Mexico. The Comision initiated studies in 1950 to investigate the possibility of generating electric power from the Cerro Prieto geothermal resource. This has resulted in the first commercial geothermal power plant in Latin America.

LOCATION

The Cerro Prieto geothermal project is located in the state of Baja California, Mexico, in the proximity of the westernmost border with the United States of America (Fig. 7-1). It lies approximately 30 km south of Mexicali, the capital of the state of Baja California, east of the Sierra de los Cucapá, on a desert plain of the Colorado River delta (Fig. 7-2). The Cerro Prieto geothermal field extends from 114°50′ to 115°48′ west longitude and 31°55′ to 32°44′ north latitude. The area is well connected with the city of Mexicali by the Sonora Baja California Railroad and by the Pascualitos-

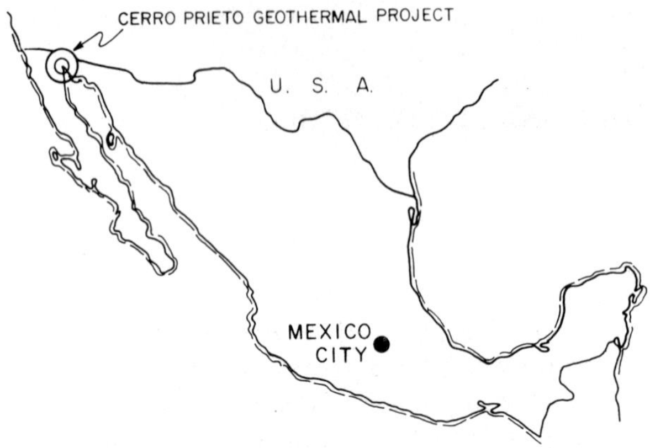

Fig. 7-1. Cerro Prieto geothermal field, Mexico.

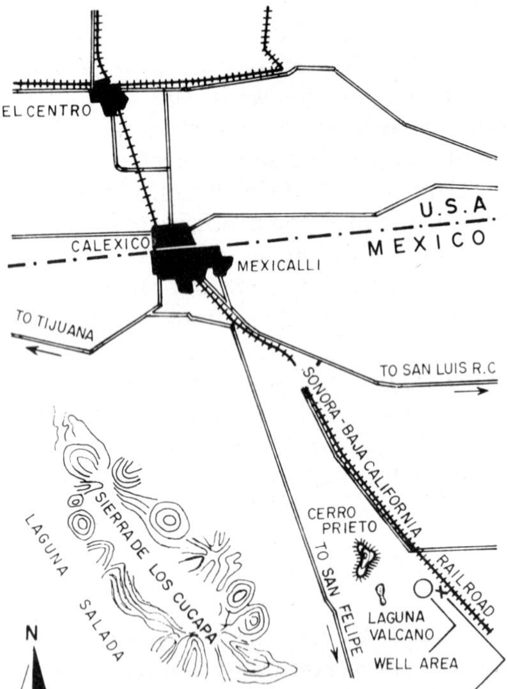

Fig. 7-2. Location details of the Cerro Prieto geothermal field.

Pescaderos state road, which in turn is connected with the Mexicali-San Luis Río Colorado Highway.

GEOTHERMAL MANIFESTATIONS

As discussed in Chapter 5, the initial exploratory stages for geothermal resources consist of studying and preparing an inventory of geothermal manifestations. Table 7-1 (taken from Banwell and Gomez Valle, 1970) summarizes the characteristics of Mexican geothermal fields. Cerro Prieto is characterized by a basaltic promontory. Around the margin of this promontory, several hydrothermal springs, sulphur deposits, geysers, fumaroles and

TABLE 7-1

Characteristics of the Mexican geothermal fields (from Banwell and Gomez Valle, 1970)

Name of field	Physiographic expression	Type of activity	Surface temperature (°C)
Cerro Prieto (88 m a.m.s.l.) field under development for 75 MW	delta valley, with surrounding farmlands	fumaroles, hot springs, mud volcanoes, lakes, and salt deposits	45—100 with high values in the wells
Los Negritos (1533 m a.m.s.l.)	caldera-like structural valley, open towards NW; flows of Tertiary and Quaternary lavas, surrounded by farmlands	mud volcanoes and bubbling lakes aligned NE-SE	36—90
Ixtlan de Los Hervores (1535 m a.m.s.l.)	E-W faulted structural valley, with flows of basalt and andesite; acid soils unfit for cultivation in the hydrothermal area	hot springs, boiling mud lakes, steam discharges, and low-pressure wells; deposits of geyserite with cermic residues	45—100
Los Humeros (1800 m a.m.s.l.)	centrally faulted caldera; explosion craters, deposits of basalts, sands, and stratified tuffs	low-pressure fumaroles, hot ground, and alteration zones	85—90
Los Azufres (2800 m a.m.s.l.)	intensely fractured double caldera; altered rhyolites and volcanic ash	superheated steam (fumaroles), large mud volcanoes, hot springs, and bubbling lakes	45—110
La Primavera (1570 m a.m.s.l.)	faulted caldera; the geothermal area is located on two N-S faults which bound a graben	superheated steam (fumaroles), alteration zones and hot ground	55—100

ponds of boiling mud are observable. These geothermal manifestations had been exploited in a rudimentary way by the local inhabitants for centuries. In the 1940's and 1950's the importance of these geothermal manifestations and the possible power potential of Cerro Prieto geothermal resource was realized. In the following years intensive geological, geophysical and geochemical investigations were carried out, and still continue, for improving the existing knowledge.

GEOLOGICAL BACKGROUND

The geology of the Cerro Prieto geothermal fields has been briefly described by Banwell and Gomez Valle (1970). The Mexicali Valley, in which the field is located, is a prolongation of the Imperial Valley, California, and it retains the physiographic features of the latter. It is constituted of thick deltaic sedimentary deposits consisting of successive horizons of sands, muds, and clays with various degrees of compaction. On the west it is terminated by the Sierra de los Cucapá (Fig. 7-2). On the east side it extends, without any interruption, to form the bed of the Colorado River. These deltaic sediments rest unconformably on intrusive granitic rocks, affected by the San Andreas/San Jacinto fault system. The granitic intrusion has been severely fractured and dislocated, resulting in numerous faults and fissures; these provide hydraulic continuity to depth, with a steam-water mixture escaping from the surface at a number of places. This causes and sustains the observed hydrothermal activity.

Cerro Prieto is a double volcanic crater of Quaternary age. It has produced basalts and pyroclastics. In its proximity and to the southwest, there is a considerable amount of fumarolic and solfataric activity. All these are aligned in a north-south direction, which probably is an extension of the known geothermal area and requires further investigation for a proper understanding.

The structural features of the Cerro Prieto area have been deduced with the help of seismic refraction and gravity surveys. A Cretaceous tectonic trench, rising in a series of stepped blocks toward the Sierra de los Cucapá and descending abruptly east of Cerro Prieto, with a displacement of the order of 2900 m, has been inferred. A comparison of detailed gravity and seismic data, and on the basis of similar observations in the Broadlands field in New Zealand, indicates that the observed increase in density in the eastern part of the area could be attributed to the deposition of minerals in the upper part of the hydrothermal system.

RESULTS OF PRELIMINARY EXPLORATION

In 1960, the first exploratory drilling to a depth of 750 m was carried out and promising geothermal gradients were discovered. After a while, a more ambitious drilling program was undertaken. In an area of 10 km^2, 42 holes were drilled. Of these, 22 produced a mixture of steam and hot water. The depth of these bore holes ranged from 900 to 2000 m. The steam and hot water were found to be produced from a horizon lying at between 650 and 900 m depth. The pressures encountered ranged from 200 to 400 psi at the wellhead and the temperatures from 250 to 340°C.

In the following years, exploration was concentrated, in general, to the southeast. Considering the observed geothermal manifestations in the north and south, it was concluded that the field had considerably more geothermal power production potential than was visualized earlier. A well, placed on the top of a buried granitic block, produced 200 tons of steam and water per hour, the wellhead pressure being 275 psi. The temperature encountered was 210°C and the producing zone was estimated to lie between 600 and 900 m depth. A second producing zone between 2400 and 2640 m was inferred near the contact with the granitic basement. It was observed that in the wells drilled, the temperature and pressure remained the same after casing. It was estimated that adequate steam and hot water would be available to operate a 75-MW power plant. The Comision Federal de Electricidad decided to install a power plant of 75 MW capacity to begin its operation in 1971. It was also estimated that in the near future it should be possible to double this power production capacity.

GEOCHEMICAL SURVEYS

The Cerro Prieto area is very arid, with an average annual rain fall of 13 cm, and the elevation is only about 10 m above the mean sea level. Consequently, the water balance is controlled mostly by evaporation. The observable surface runoff is very small. Numerous chemical analyses of samples from natural activity observable at the surface, as well as from boreholes are available. Molina and Banwell (1970) have summarized these data. Isochemical maps have been prepared using both these sets of data and they exhibit well-defined variations across the field and correlate well with other characteristics, such as temperatures measured in the boreholes. Similarly, well-defined patterns have been revealed on the plots of certain surface water constituents. These patterns are centered around localized points of minimum and maximum values, indicating that thermal water surfaces at certain discrete sources. These points are also found to be aligned along faults which have been identified from other evidence. Steam, hot water and gases, rising from the deep reservoir, undergo chemical modification and separation as they approach the surface. Similarly, the geothermal water, as it spreads over the surface from the source undergoes further changes as a consequence of mixing, evaporation and reaction with atmospheric oxygen.

Fig. 7-3, from Molina and Banwell (1970), shows the results of atomic ratio determinations for Mexican geothermal fields. On chloride-boron-bicarbonate as well as chloride-sulphate-bicarbonate plots, the Cerro Prieto field separates very sharply from the rest (7-3A,B). Cerro Prieto samples are also distinguished by a relatively higher proportion of sodium and potassium relative to silica (Fig. 7-3C). Temperatures inferred on the basis of chemical analyses of samples from drill holes in the Cerro Prieto region show wide variations which are consistent with the wide range of temperatures and chemical conditions observed over the field. However, in general, geochemical analyses showed favorable results and temperatures of about 350°C could be inferred from the samples from certain boreholes.

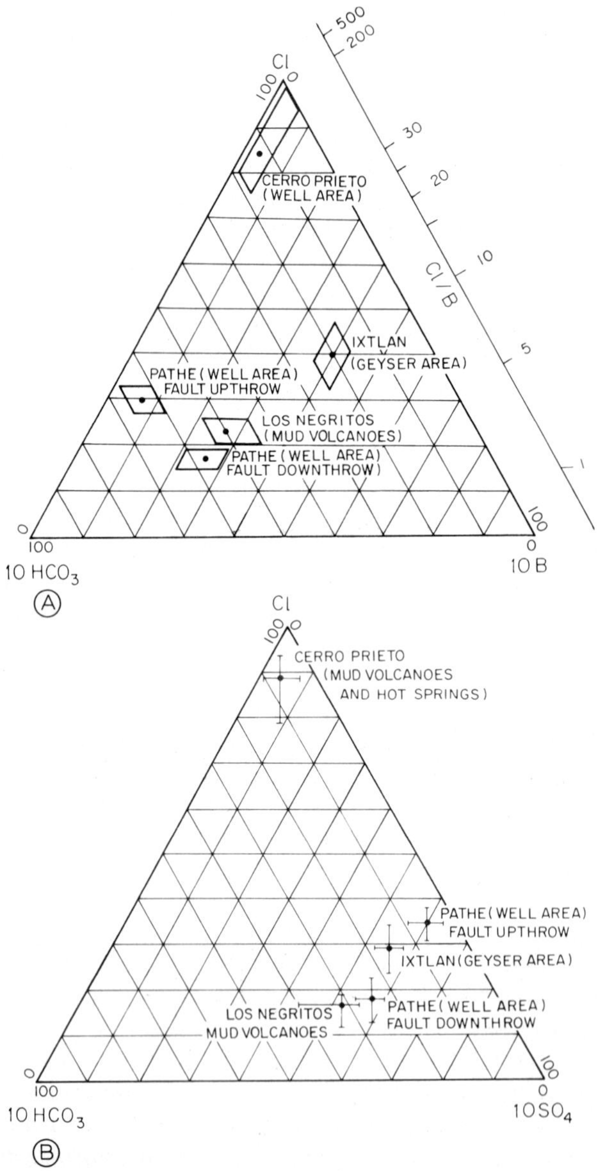

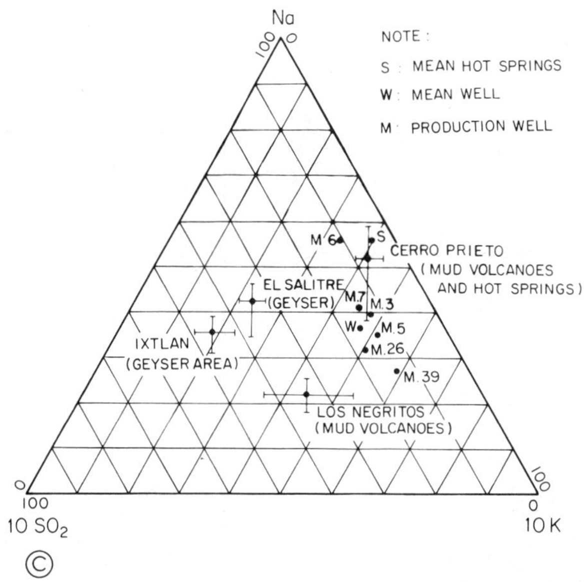

Fig. 7-3. Geochemistry of geothermal fields in Mexico (modified from Molina and Banwell, 1970). Atomic ratios. A. Cl : HCO_3 : B. B. Cl : HCO_3 : SO_4. C. Na : SO_2 : K.

ELECTRICAL SURVEYS

The electrical surveys carried out at Cerro Prieto are summarized by Garcia (1976). The first phase included apparent resistivity measurements at constant depth through profiling. The area was found to be characterized by low resistivity values (less than 10 Ω-m), with an area of about 70 km^2 having an apparent resistivity of less than 2 Ω-m (Fig. 7-4). The Cerro Prieto volcano is located within this zone of low resistivity.

In the second phase, Schlumberger depth soundings were carried out within a 40-km^2 zone around the presently producing area. Fig. 7-5, from Garcia (1976), shows resistivity contours at the deepest interpreted depth of 1200 m. The resistivity contours have a northwest-southeast trend, parallel to the regional fault system. Several other profiles were investigated in subsequent years. The east-west profile (Fig. 7-6) is an interesting one, showing a fault in the deepest layer. The excellent water-steam production from a well close to this fault, inferred from resistivity data, indicates that this fault is one of the most productive in the Cerro Prieto geothermal field.

In conclusion, it may be said that electrical resistivity surveys have been extremely useful in delimiting the boundaries of the Cerro Prieto geothermal field as well as in deciphering its northwest-southeast alignment. There is a good correlation between the observed low resistivity values and high production from wells. Wells located close to basement faults, inferred from

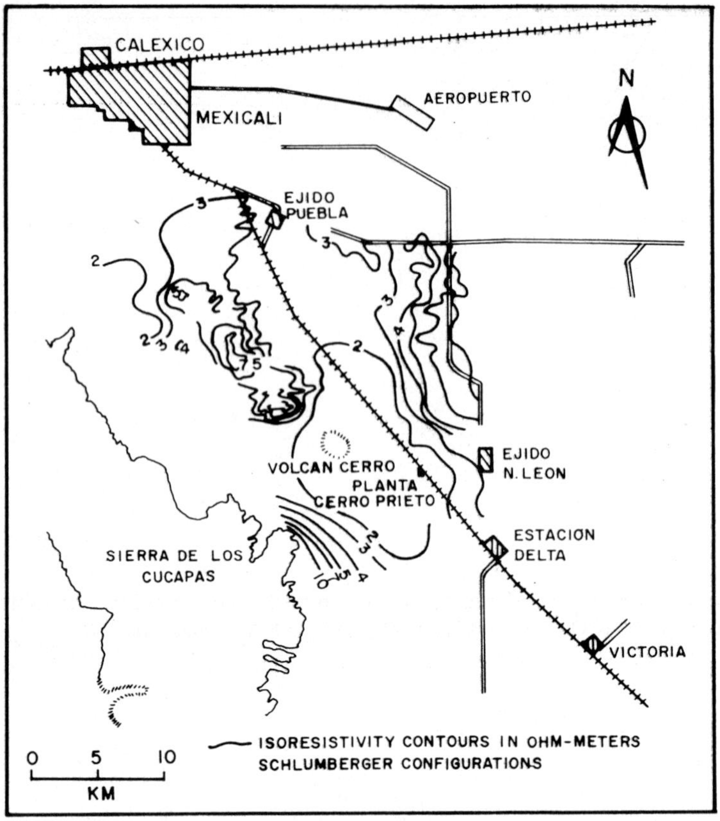

Fig. 7-4. Isoresistivity contours at Cerro Prieto (modified from Garcia, 1976).

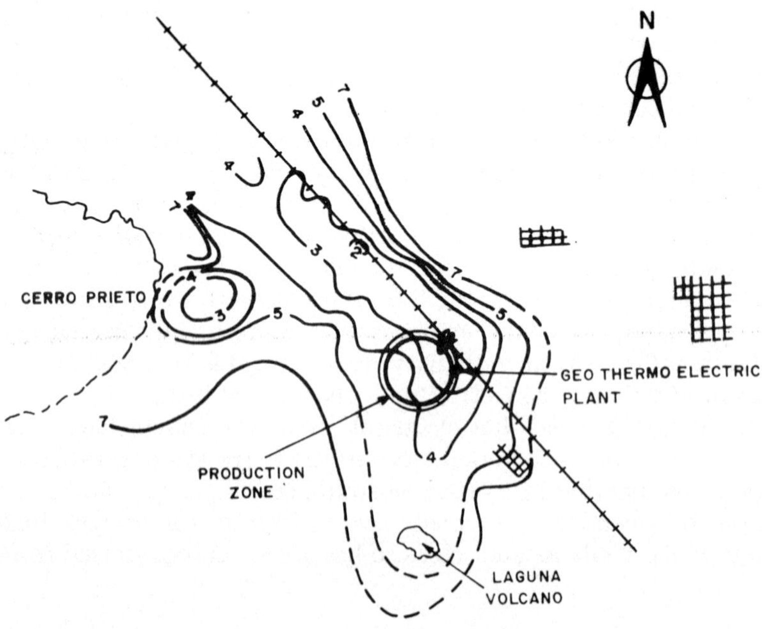

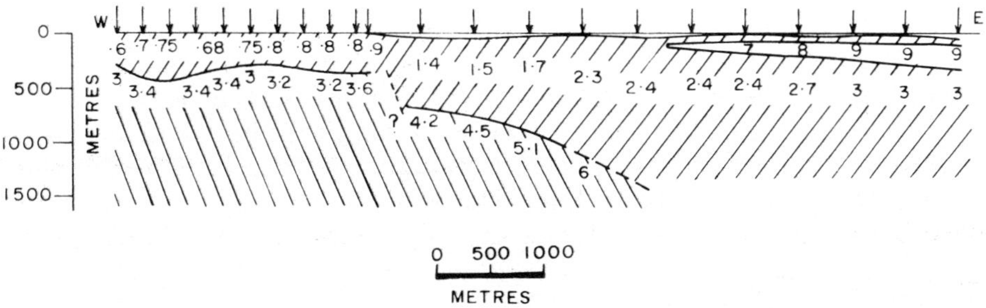

Fig. 7-6. An east-west section across Cerro Prieto. Inferred resistivities are given in ohm-meters (modified from Garcia, 1976).

resistivity data, have been particularly good in the production of geothermal fluid.

AN EAST-WEST MODEL SECTION THROUGH CERRO PRIETO

As a result of integrated geological, geochemical and geophysical investigations and inspection of cores obtained during the drilling of deep wells, the structure of the Cerro Prieto geothermal field is pretty well understood; a schematic east-west section, across it is shown in Fig. 7-7. The granitic base-

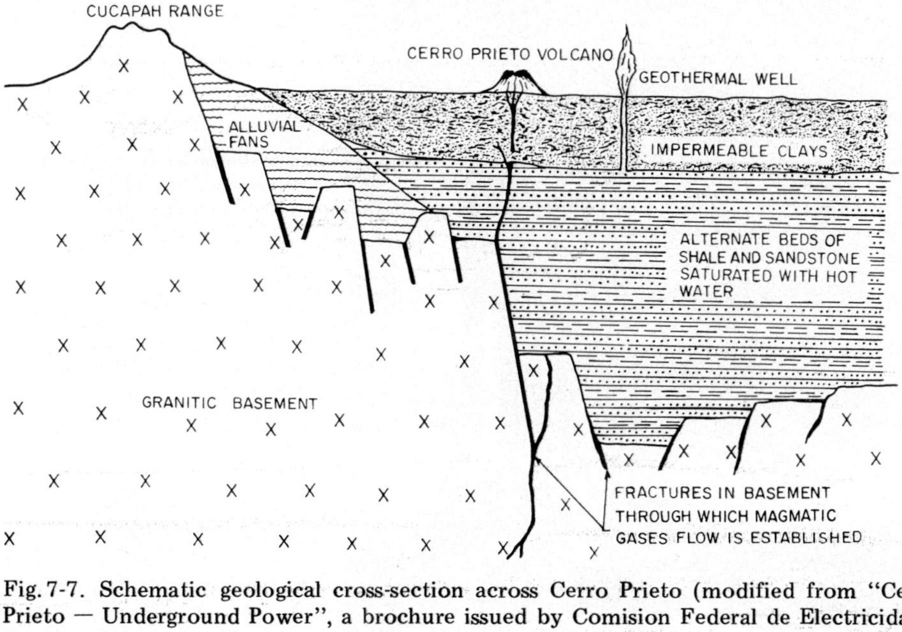

Fig. 7-7. Schematic geological cross-section across Cerro Prieto (modified from "Cerro Prieto — Underground Power", a brochure issued by Comision Federal de Electricidad). For details, see text.

Fig. 7-5. Isoresistivity contours at a depth of 1200 m at Cerro Prieto (modified from Garcia, 1976).

ment lies approximately at a depth of 2500 m; it is fractured and faulted and shoals towards the west. Tectonically the zone is believed to be related to the San Andreas system of the United States. The basement is overlain by alternate beds of shales and sandstones. These sandstone horizons, saturated with super-heated steam and water, act as the reservoir from which geothermal fluid is produced by drilling. The surficial, impervious deltaic sediments, approximately 70 m in thickness, keep the geothermal fluids sealed.

DRILLING

As discussed in Chapter 6, drilling of geothermal wells differs considerably from drilling for oil. At Cerro Prieto, after having chosen a drilling site, the ground around it is compacted to ensure its stability. Next a trench is dug, to a depth of 2—3 m, for the drilling cellar, and the rotary drilling equipment

Fig. 7-8. A rotary rig at Cerro Prieto (photograph by the author).

is installed. A typical rotary drilling rig and accessories, used at Cerro Prieto, are shown in Fig. 7-8.

The drill hole diameter decreases with depth. At Cerro Prieto, drilling typically starts with a 22″ diameter hole. As the depth increases, it is reduced to 15″, and at the depth of 1000 m, it is further reduced to $10\frac{5}{8}$″. This diameter is normally retained to the bottom of the hole, which usually is 1400 m. The drilling fluid consists of bentonite slurry with several additives to withstand the high temperatures encountered. A cooling tower is installed to cool the circulating slurry.

The drilling fluid is often lost through fissures and fractures encountered during the course of drilling. In the event that these fissures and fractures are unproductive, they are sealed with vegetable fibers, mica flakes, celophane chips, or, in some cases by cement. If these fissures lie in an adequately productive zone, drilling is terminated and the hole is considered to have been completed. With the progress of drilling, casing is provided. At Cerro Prieto, seamless steel pipes of 16″ diameter are used for anchoring, $11\frac{3}{4}$″ diameter pipes for lining and $7\frac{5}{8}$″ diameter pipes for production. The production casing is slotted. Portland cement, with additives to delay setting, is used for

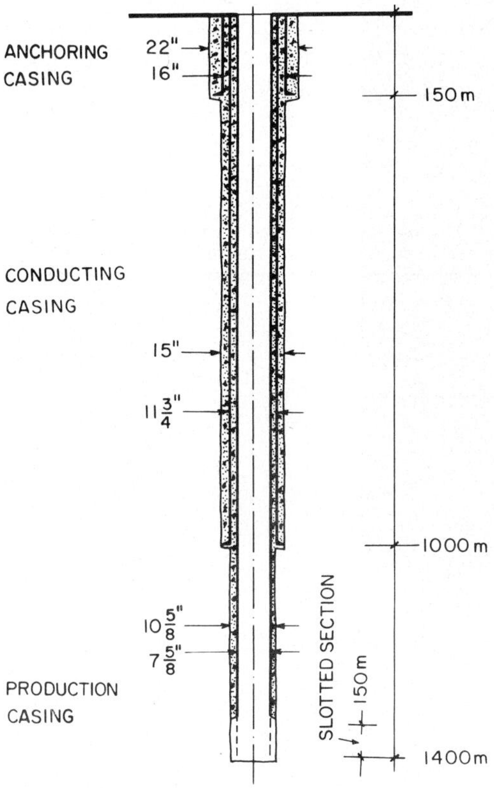

Fig. 7-9. Schematic section of a geothermal well at Cerro Prieto. For details, see text.

binding the casing with the borehole. Fig. 7-9 schematically shows a typical borehole.

Geothermal fluid producing horizons are detected by means of electrical, thermal and sonic logging. During drilling, the necessary safety equipment is installed at the anchor pipe to keep the well under control. This essentially consists of a blow-out preventor, which automatically seals the well when pressures become dangerously high. After termination of drilling, the safety equipment is withdrawn and a valve tree (Fig. 7-10) is installed on the pipe head. Geothermal fluid is taken out through different ducts from the valve tree.

Fig. 7-10. A typical valve tree at Cerro Prieto (photograph by the author).

PRODUCTION

Cerro Prieto geothermal wells produce a mixture of hot water and steam. It is necessary to separate them and send the dry steam to the turbine of the geothermoelectric power plant. This is achieved by passing the well output through a centrifugal separator (Fig. 7-11). The centrifugal force makes the water adhere to the walls of the separator, and it finally descends to the bottom, while the steam rises in the separator and is removed through a pipe. The separated water is discharged into the silencers and later into the drainage channels.

The silencers are useful in reducing the noise produced by discharging the geothermal fluids at a high pressure into the atmosphere. At Cerro Prieto, horizontal as well as twin-barrel vertical silencers are used. The horizontal silencer uses a series of pipes with increasing diameters. As the fluid passes through it, the flow speed gradually decreases and the fluid is discharged at a much diminished noise level. The twin-barrel vertical silencer consists of two vertical wooden pipes resting on a concrete structure. The discharge pipe from the separator releases fluids at the concrete structure. With the release of pressure, steam is formed, and water and steam become separated due to centrifugal action, similar to that in the separator. Eventually, steam is released at a low pressure, and consequently with reduced noise, to the atmosphere and the water to the drainage channel.

Fig. 7-11. A centrifugal separator at Cerro Prieto (photograph by the author).

TRANSPORTATION OF STEAM

Hundreds of meters of pipeline are used to bring the steam from the separator to the electric power plant turbine. Several steam traps are installed to remove water formed by condensation of steam, since the steam must be dry when it enters the turbines. As a measure of additional precaution, the steam is passed through a moisture absorber before entering the turbines. All production wells are connected through four branches of pipelines. Pipeline joints are so constructed that they are capable of absorbing the mechanical stresses generated by thermal expansion of the pipes. At the power plant, the four branches of pipelines are connected to two steam collectors. The total length of pipeline, with a pipe diameter of 16—22 inches, is in excess of 6 km. The secondary pipelines of small diameters add up to several more kilometers. Fig. 7-12 is a schematic representation of the wells and pipelines at the Cerro Prieto.

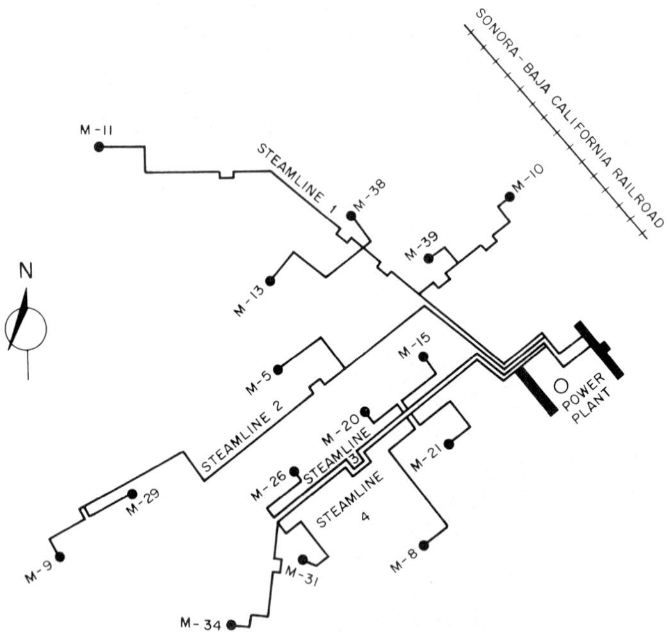

Fig. 7-12. A sketch of the location of wells and steam transportation pipes (modified from "Cerro Prieto — Underground Power", a brochure issued by Comision Federal de Electricidad).

THE POWER PLANT

The Cerro Prieto geothermoelectric power plant is a prestigeous project of Comision Federal de Electricidad, Mexico (Fig. 7-13). The infrastructure of the plant is quite conventional. The structure, made of galvanized steel, covers an area of $60 \times 14.5\ m^2$ and has a height of 30 m. Figs. 7-14 and 7-15 give general views of the power house. A schematic layout of the power house and the accessories is given in Fig. 7-16. The turbogenerator and the

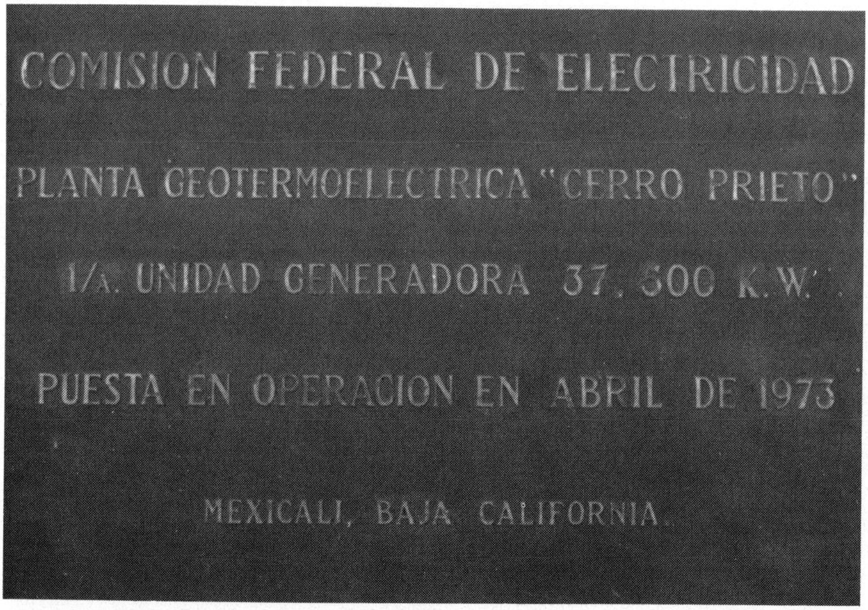

Fig. 7-13. Comision Federal de Electricidad.

Fig. 7-14. A general view of the power house at Cerro Prieto (photograph by the author).

Fig. 7-15. Another view of the power house at Cerro Prieto (photograph by the author).

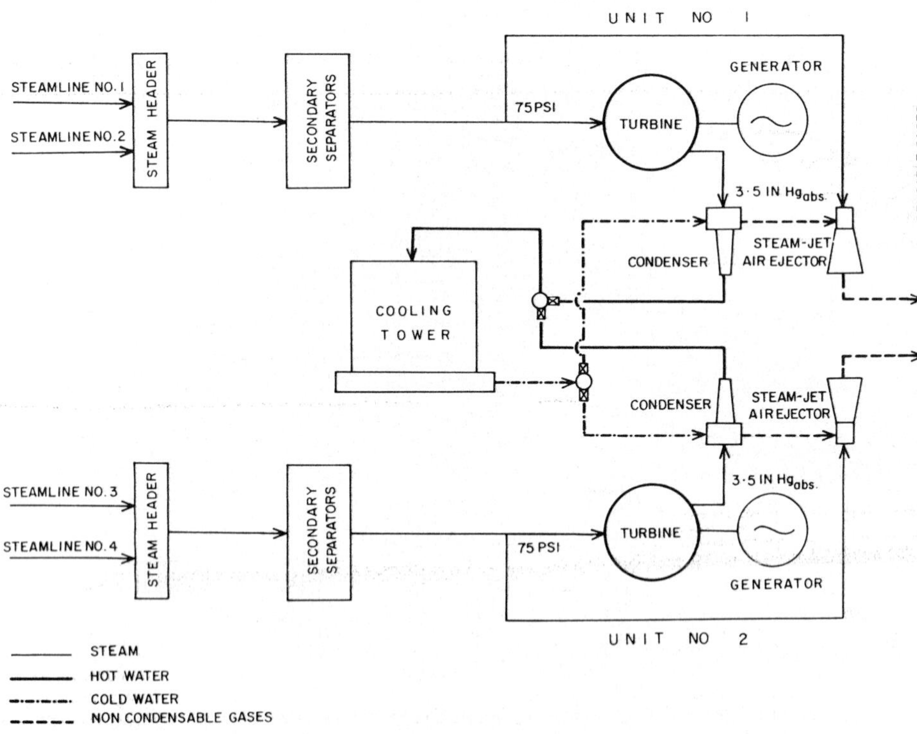

Fig. 7-16. Layout of the power house at Cerro Prieto (modified from "Cerro Prieto — Underground Power", a brochure issued by Comision Federal de Electricidad).

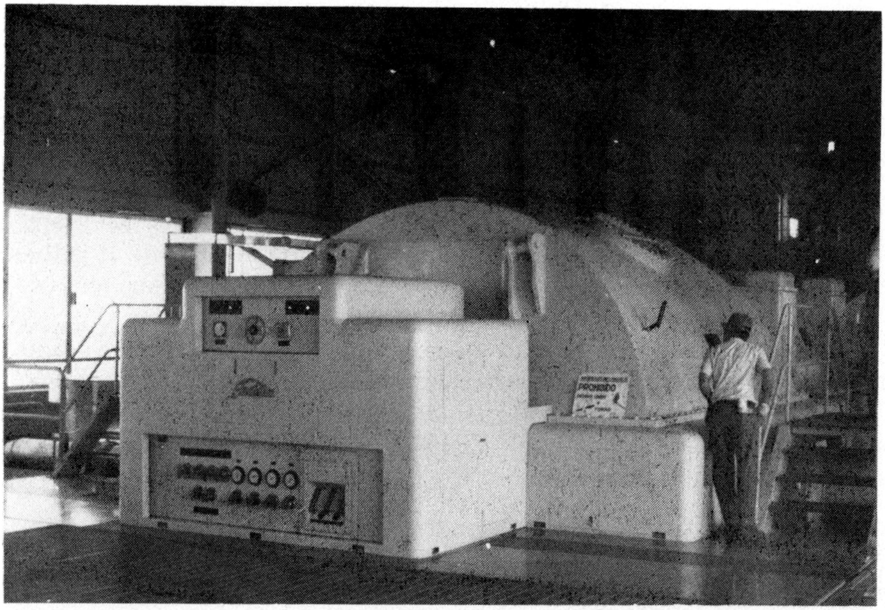

Fig. 7-17. A turbogenerator at the Cerro Prieto power house (photograph by the author).

Fig. 7-18. A view of the cooling tower at Cerro Prieto (photograph by the author).

auxiliary equipment are installed inside, while the condensors, because of their size, are installed outside the engine house. The power house is a double-story structure with the turbogenerators, oil and hydrogen systems and several laboratories, operational and administrative offices. All turbogenerators are built with special metal alloy to reduce corrosion and erosion. Fig. 7-17 shows a turbine.

The Cerro Prieto geothermal field is located in a desert and the scarcely available water does not meet the requirement of the power plant. It is therefore necessary to recover water formed by condensation of steam and use it for cooling purposes. This is achieved through a cooling tower (Fig. 7-18) built for the purpose. The cooling tower is huge ($146 \times 25 \times 18$ m^3) and is capable of cooling 27,000 m^3 of water per hour to 16°C.

The substation and main transformer are located close to the power house. This set-up is quite conventional. Fig. 7-19 is a view of the substation and transformer and Fig. 7-20 is a view of the power plant control room.

Fig. 7-19. A view of the substation and transformer at Cerro Prieto (photograph by the author).

Fig. 7-20. A view of the control room at Cerro Prieto (photograph by the author).

COSTS

According to "Cerro Prieto — Underground Power", a brochure released by the Comision Federal de Electricidad, Mexico, the expenditure up to the stage of commissioning the 75-MW unit had been $14,400,000. Of this sum, $7.2 million were spent on the power plant equipment, $2.1 million on drilling and the rest of the money was spent on labor and overhead costs. The annual expenses for running the power plant are about $1.5 million.

FURTHER DEVELOPMENT

As reported by Tolivia (1976), several studies have been made to determine whether power production at the Cerro Prieto should be increased from 75 to 150 MW. These investigations include the drilling of new wells outside the current production area, several hydraulic and thermodynamic measurements in the producing wells, and direct and indirect measurements of permeability and porosity of the reservoir rocks as well as consideration of several field models. The potential of the reservoir has been estimated for two basic models. The first assumes a thermally and hydraulically sealed reservoir, the second takes into account the known thermal and hydraulic recharge. For the first case, it is estimated that a 150-MW unit would be kept supplied for at least 33 years, and the life time would increase to 90 years when the known thermal and hydraulic recharge is taken into consideration. Therefore, installation of another unit increasing the power generation to 150 MW is warranted.

Chapter 8

WORLD-WIDE STATUS OF GEOTHERMAL RESOURCE UTILIZATION

The geothermal resource utilization can broadly be classified into two categories: (1) electric power generation, and (2) non-electrical use.

ELECTRIC POWER GENERATION

Various aspects of geothermal electric power generation have been discussed and summarized by Wood (1973), and Armstead (1976). Commercially available geothermal power plants are distinctive, primarily because they use low-pressure steam turbines for driving the generators. After passing through the turbine, the steam is mixed with cold water in the condensor and converted into hot water. The hot water is then taken to evaporating cooling towers, where almost three-fourths is evaporated and the remaining cold water is either recirculated through the condensor or reinjected into the reservoir. This sequence of power generation applies to both vapor-dominated and hot-water-dominated systems (Figs. 8-1 and 8-2). As listed in the document entitled "Energy Alternatives: A Comparative Analysis" (1975), a geothermal plant has the following distinctive features:

(1) In a geothermal power plant, there is no combustion of any fuel.

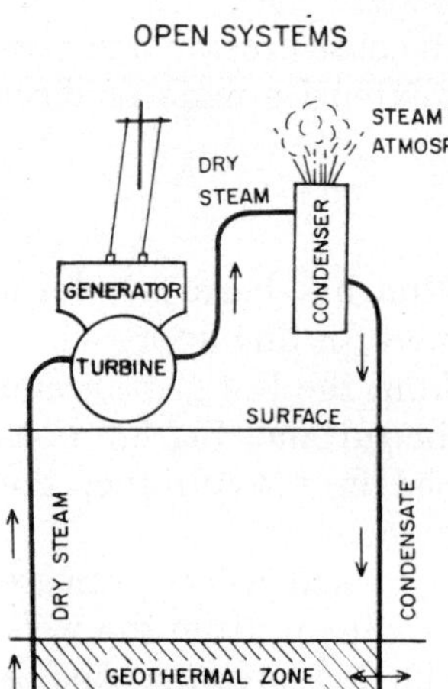

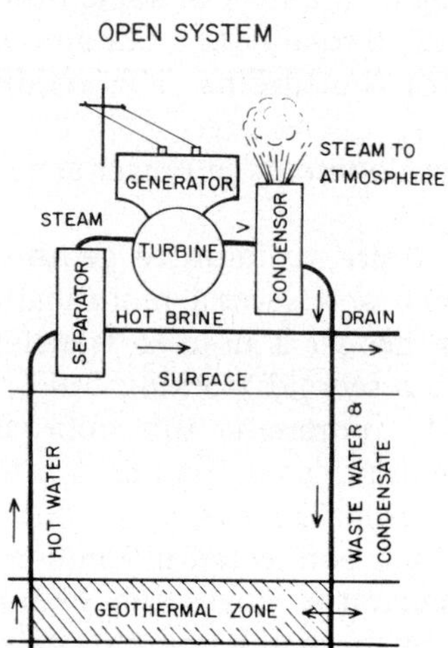

Fig. 8-1. Working principle of a dry-steam-type geothermal power plant.

Fig. 8-2. Working principle of a hot-water-type geothermal power plant.

(2) Low temperature and pressure of the steam result in low efficiency. The overall plant efficiency in geothermal power production is approximately 15% compared to 35—38% for fossil-fueled power plants.

(3) Long and complicated start-up procedures render the geothermal power units more suitable as base-load units than as peak-load units.

(4) Since transportation of steam over long distances through pipelines results in substantial loss of heat, geothermal power plants need to be installed close to the site of steam or hot water production. They are also relatively much smaller in their production capacity.

(5) In general, a 110-MW station requires two million pounds of steam per hour, that is the combined output of 14 wells, each producing 150,000 pounds of steam per hour.

(6) Steam and cooling water mix directly in the contact condensers.

(7) Usually no external make-up water is required for cooling. The cooling tower evaporation rate is exceeded by the steam flow to the turbines. Thus, condensed exhaust is used as cooling tower make-up water.

(8) Non-condensable gases are released into the atmosphere from the condenser and from the cooling tower.

(9) At The Geysers, California, 75—80% of the condensed steam evaporates in the cooling tower and 20—25% is reinjected into the steam-bearing formations.

(10) The water or brine in the hot water system (Fig. 8-2) is passed through a separator, which draws off steam to run the turbine, and then routed to reinjection wells. At times, additional water from cooling towers also needs to be disposed of. This is done either through re-injection or diverting it to a river or some other suitable body of water.

(11) The steam has a fair amount of minerals which cause erosion and corrosion of the turbine. This requires continuous and extensive maintenance.

New developments in power generation systems

The desire to improve geothermal power generation has been felt for a long time and several modifications are being adopted. In one approach, a heat exchanger is used to transfer the heat energy from the hot geothermal fluid to a second working fluid, which is fed into the turbine. Fig. 8-3 conceptually represents this approach. After being used in the turbine, the working fluid is pumped back into the circulation system.

In another approach for hot water systems, a high- and a low-pressure turbine are connected in tandem, as shown in Fig. 8-4. Steam from the wellhead separator drives the high-pressure turbine. The low-pressure one is driven by both the high-pressure turbine exhaust and the steam created by flashing the reject from the wellhead separator.

In yet another approach, the development of impulse turbines, a high-velocity jet of water impinges on vanes or buckets at the wheel periphery. This high-velocity jet of water is obtained from the geothermal water-steam mixture by expanding it through a converging-diverging nozzle to a low pressure and high velocity. This system is termed total fluid flow (Smith et al., 1973). A variation of the impulse turbine is the helical screw expander, which is also under active development.

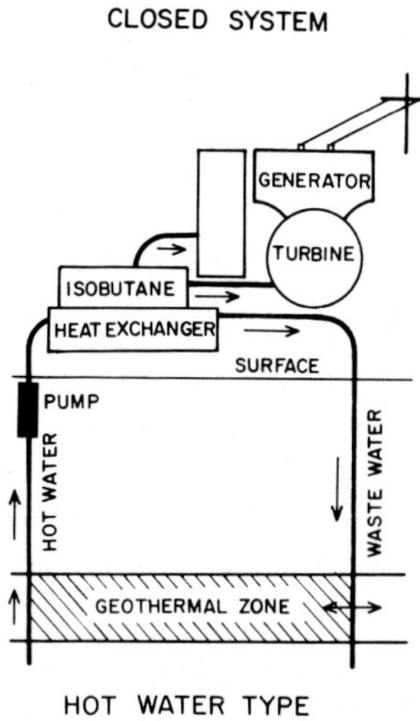

Fig. 8-3. Working principle of a hot-water-type geothermal power plant under development. For details, see text.

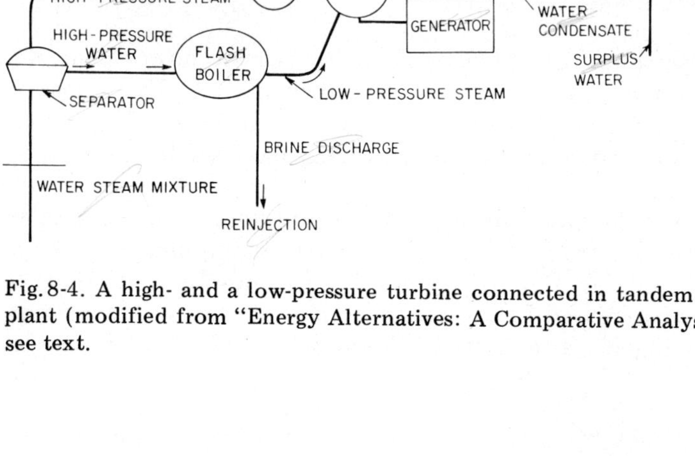

Fig. 8-4. A high- and a low-pressure turbine connected in tandem in a geothermal power plant (modified from "Energy Alternatives: A Comparative Analysis", 1975). For details, see text.

Energy efficiency in geothermal electric power generation

The power generation step is more efficient in a vapor-dominated geothermal system, since the steam enters the turbine at a higher temperature and pressure, compared to a hot-water-dominated system. Similarly, the power generation step in a heat exchanger, or impulse turbine is less efficient because of lower temperatures and pressures. However, it is likely that the total efficiency in a hot water system may be equal to, or even higher than a vapor-dominated system because of the heat content of the water used. Table 8-1, taken from the document "Energy Alternatives: A Comparative Analysis" (1975), shows the system efficiency from the wellhead through electric power generation (inclusive of reservoir recovery efficiency) for four geothermal energy systems. It may be noted from this table that the total fluid flow impulse turbine system has the best efficiency. However, this estimate is based on theoretical and pilot plant results and is still to be proved commercially. Using steam from a vapor-dominated geothermal reservoir is currently the most efficient system.

TABLE 8-1

System efficiency: wellhead through electric power generator, inclusive of reservoir recovery efficiency (modified from "Energy Alternatives: A Comparative Analysis, 1975)

Type	Percent efficiency
Vapor-dominated reservoir*	14.7
Hot-water-dominated reservoir** (flushed steam)	11.0
Binary fluid type**	11.0
Total flow impulse turbine**	18.0

*Data from The Geysers, California (Teknekron, 1973).
**Data from Austin et al. (1973).

Environmental considerations

Estimates, from Reed and Campbell (1976), for electric power plant air-pollutant emissions and the cooling tower effluents at The Geysers, California, are given in Tables 8-2 and 8-3. Hot-water-dominated reservoirs usually have a higher percentage of non-condensable gas fractions than vapor-dominated ones.

Hydrogen sulphide is the most troublesome air pollutant. According to an estimate, the total release from The Geyser-type system would range from 100,000 to 215,000 pounds of H_2S per day for a 1000-MW unit (Battelle Columbus and Pacific Northwest Laboratories, 1973). For hot water systems, H_2S release is much higher. The Battelle Columbus and Pacific Northwest

TABLE 8-2

Composition of flow from the turbine gas ejectors at The Geysers field* (Reed and Campbell, 1976)

Constituent	Average concentration (wt. %)	Average flow from 110-MW unit (kg/hr)
Carbon dioxide	59.9	2500
Nitrogen**	25.8	1100
Oxygen**	8.4	350
Methane	3.8	160
Hydrogen sulphide	1.2	50
Hydrogen	0.9	38

*From measurements of nine generating units from 1967 through 1974, Pacific Gas and Electric Company.
**Nitrogen and oxygen are from air entering at the turbine.

TABLE 8-3

Estimated output of pollutants from plant cooling towers at The Geysers field* (Reed and Campbell, 1976)

Constituent	Average concentration (mg/kg)	Average flow from 100-MW unit** (kg/hr)
Ammonia	240	160
Carbon dioxide	210	140
Hydrogen sulphide	200	130
Hydrogen	12	8

*Pacific Gas and Electric Company
**Based on water vapor output of 658,000 kg/hr.

Laboratories (1973) have estimated that H_2S release at Cerro Prieto, Mexico, is approximately 10 times greater than at The Geysers. This exceeds the amount of H_2S released from burning high-sulphur fuels in fossil-fuel-fired power generators. Allen and McCluer (1976) have summarized the hydrogen sulfide emission abatement program at The Geysers and the actions under way to reduce it. On average, The Geysers steam has 223 ppm H_2S by weight. This H_2S, after passing through the turbines, leaves the power-generating units through the gas ejector system and by air stripping in the cooling towers. Allen and McCluer (1976) point out that the hydrogen sulphide dissolved in the cooling water can be controlled by using an oxidation catalyst, such as an iron salt, so that it is converted to sulphur before entering the cooling tower. The H_2S in the gas ejector system could be burned to form sulphur dioxide and scrubbed directly into the circulating water for reinjection into the steam field.

Other chemicals, such as mercury, radon, ammonia, boron and fluoride drift from the cooling towers and rain out. The severity of these pollutants varies from one site to another.

Heat emitted to the atmosphere via a cooling tower is approximately 2.5 times greater in a hot-water-type reservoir than in a vapor-dominated one.

NON-ELECTRICAL USE

Non-electric use of geothermal resources has been discussed in detail in documents edited by Howard (1975), and Lienau and Lund (1974). Armstead (1976) and Barr (1976) have summarized several aspects of geothermal energy utilization in space and process heating, and various other single- and multi-purpose usages.

Until recently, there was a tendency to think of geothermal development primarily in terms of electric power production only. This is essentially an inefficient process due to inescapable thermodynamic restraints. With the lowering of the temperature of the heat source, the efficiency of power generation goes down. Quite contrary to this situation, the direct use of geothermal heat for space heating, industrial processes, domestic hot water supply, etc., are much more efficient, since the losses incurred are not imposed by the laws of thermodynamics. In these usages, the heat is lost primarily due to inadequate insulation and terminal temperature differences in heat exchangers and drains. Table 8-4 gives estimates of the efficiency of utilization of geothermal energy in the Námafjall area of Iceland.

The thermal distribution system, which is an assembly of piping with varying diameters of pipes depending upon the flow rates, is the most important component of the set-up for non-electric use of geothermal energy. Other equipment used includes pumps, regulators, valves, meters, expansion joints and the controls necessary for reliable operation of the network. To prevent excessive heat loss and temperature drop in the fluid, the system needs to be thermally insulated. There is a continuous decline in the fluid pressure along the pipelines due to viscous friction losses. Changes in elevation along the pipeline effect the hydrostatic head. Therefore, to prevent the

TABLE 8-4

Efficiency of geothermal energy utilization in non-electric power production at Námafjall, Iceland

Purpose	Percent efficiency
District heating, using both steam and water	about 80
Industrial heating, using steam only	about 35

pressure from dropping below a value which would result in local boiling, and to maintain sufficient flow throughout the entire network, pumps are essential. Surge tanks are required to prevent sudden and potentially damaging changes in the pressure due to sudden changes or surges in the flow.

Impurities, such as hydrogen sulphides, carbon dioxide, chlorides, silica, bicarbonates and entrained sand particles cause corrosion, erosion and scaling. Chemical corrosion and/or mechanical erosion can cause pipe failure, whereas scaling increases pumping requirements as well as disrupting the operation of equipment such as heat exchangers. Depending upon the severity of impurities, it may be necessary to provide special material for lining the pipes.

Necessary allowance also needs to be made to accommodate axial expansion of metallic pipes due to high temperatures of the fluid. If adequate precaution is not taken, excessive stresses caused by axial expansion could cause failure in the pipe, supports, joints or anchors. Conventional expansion devices, such as slip joints, bellows, U- or Z-shaped expansion joints are found to provide adequate protection.

Various methods are in use to provide the necessary insulation for the piping. These range from putting the prefabricated pipe within another pipe with insulation in between, to placing the pipe in concrete trenches. Insulation material, such as calcium silicate, polyurethane, mineral wool, and fiber glass, are also used to reduce conduction losses. After considering the type of soil, local water table, etc., through which piping passes, suitable installation and insulation material should be chosen.

Very often geothermal fluid needs to be transmitted several kilometers from the site of its production. In such cases, it is necessary to estimate the temperature drop as a function of distance of transmission, flow rate, fluid temperature, etc. Fig. 8-5 (Howard, 1975) shows temperature drop as a

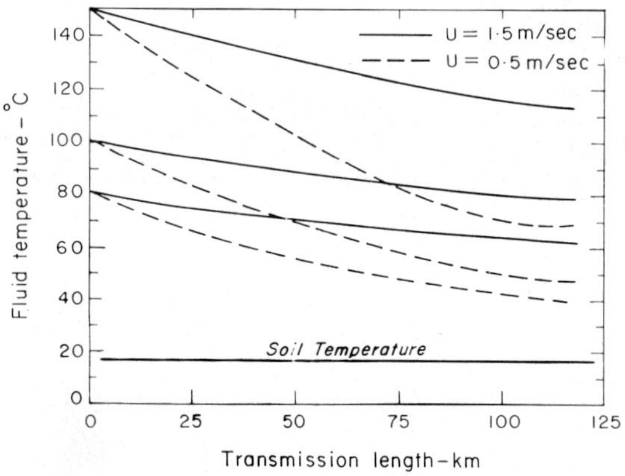

Fig. 8-5. Diagram showing effect of distance of transmission on fluid temperature in an 18" diameter pipe insulated with 2" of urethane. The soil temperature is 16°C and the flow rates considered are 1.5 and 0.5 m/s (modified from Howard, 1975).

function of distance in an 18″ diameter pipe covered with 2″ of urethane insulation. For a soil temperature of 16°C and flow rate of 1.5 m/s, the fluid temperature drops from 100 to 80°C after moving 100 km.

As listed in Howard (1975), the wide variety of non-electrical uses of geothermal energy include the following:

(1) Industrial: (a) chemical production, (b) pulp treatment, (c) mining, (d) water desalination, (e) water distillation, (f) mineral recovery from hydrothermal fluid, and (g) waste treatment and disposal.

(2) Residential and commercial: (a) space heating and cooling, (b) water (potable, hot/cold utility, etc.), (c) refrigeration, (d) waste disposal and bi-conversion, etc., and (e) deicing.

(3) Agriculture and related areas: (a) animal husbandry, (b) aquatic farming, (c) greenhouses, hydroponics, heated soil, etc., and (d) agricultural product processing such as drying, fermentation, waste disposal and conversion, and canning.

WORLD-WIDE UTILIZATION OF GEOTHERMAL ENERGY

Table 8-5, from Muffler (1976b), gives the status of geothermal electric generation in 1975. Table 8-6, based on the data in Howard (1975), gives non-electrical use of geothermal energy in different countries.

TABLE 8-5

World geothermal generating capacity (in MW) in 1975 (modified from Muffler, 1976b)

Country	Field	Operating	Under construction
United States	The Geysers	502	216
Italy	Larderello	380.6	
	Travale	15	
	Monte Amiata	22	
New Zealand	Wairakei	192	
	Kawerau	10	
Japan	Matsukawa	22	
	Otake	13	
	Onuma	10	
	Onikobe	25	
	Hatchobaru		50
	Takinoue		50
Mexico	Pathé	3.5	
	Cerro Prieto	75	75
El Salvador	Ahuachapan		90
Iceland	Námafjall	2.5	
	Krafla		55
Philippines	Tiwi		100
Soviet Union	Pauzhetsk	5	
	Paratunka	0.7	
Turkey	Kizildere	0.5	2.5
Total		1278.8	638.5

TABLE 8-6

Non-electric utilization of geothermal energy based on the data given by Howard (1975)

Country	Application	Associated power (MW)
U.S.S.R.	District heating, industrial use, greenhouses and animal husbandry	1126
Hungary	District heating, greenhouses and animal husbandry	363
Iceland	District heating, greenhouses, drying seaweed, aquaculture and industrial use	264
New Zealand	District heating, greenhouses, animal husbandry and pulp and paper industry	125
Japan	District heating, greenhouses, animal husbandry, eel breeding and rice drying	33
Italy	Greenhouses and chemical industry	24
U.S.A.	District heating, greenhouses and production of dry ice from CO_2	16
Philippines	Salt plant and grain drying	5
France	District heating	4
Turkey	Greenhouses	1

United States of America. The status of geothermal development in the United States was recently summarized by Koeing et al. (1976). Only at The Geysers, California, is electricity generated from geothermal energy. The Geysers geothermal field, at the moment, is producing the maximum electric power from geothermal energy of anywhere in the world. Fig. 8-6 shows a general view of The Geysers. There, the production has increased from 78 MW in 1970 to 502 MW in 1975. At The Geysers, a yearly increase of at least 100 MW is planned for next several years. Fig. 8-7, modified from Rex (1978), exhibits the installed, under construction and planned power generation capacity at The Geysers. As listed by Muffler (1976b), significant geothermal discoveries in the recent years have been made at the Valles Caldera, New Mexico; Roosevelt Hot Springs, Utah; Carson Desert, Nevada; and several areas such as Heber, East Mesa and Brawley in the Imperial Valley, California. The use of geothermal heat for space heating is increasing, but still is on a small scale, primarily being at Klamath Falls, Oregon and at Boise, Idaho. Fig. 8-8, from Muffler (1976b) shows the locations of geothermal drilling in the United States and northern Mexico. Table 8-7 summarizes geothermal drilling carried out during 1977 in the western states of the United States of America. Muffler attributes lack of extensive develop-

Fig. 8-6. A view of The Geysers geothermal area in California (photograph by the author).

ment of geothermal resources in the United States to institutional problems. These include ownership considerations (surface versus subsurface), leasing delays, legal definition of geothermal resources, uncertainties about the tax status, and environmental litigation.

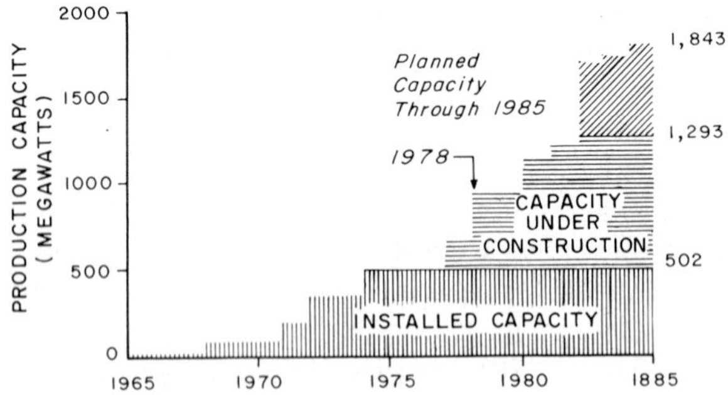

Fig. 8-7. The installed, under construction and planned power generation capacity at The Geysers (modified from Rex, 1978).

TABLE 8-7

Summary of geothermal drilling in the western states of the U.S.A. (modified from Smith et al., 1978)

State, region	Area	Wells	Producible	Suspended	Abandoned	Injection	Observation	Footage
California								
Imperial Valley	East Mesa	5	3	0	0	2	0	30,581
	Brawley	2	1	1	0	0	0	17,539
	Westmoreland	1	1	0	0	0	0	1938
The Geysers	Main Geysers	14	11	2	1	0	0	116,641
	Southeast Geysers	13	8	2	3	0	0	101,840
	Northwest Geysers	2	0	1	1	0	0	19,965
	North Geysers	1	0	1	0	0	0	4113
	Howard Hot Springs	1	0	1	0	0	0	7376
	Thurston Lake	1	0	1	0	0	0	10,560
Inyo Co.	Coso Hot Springs	1	0	1	0	0	0	4845
Nevada								
Carson Sink	Soda Lake	1	0	0	0	0	1	4975
	Allen Springs	1	0	0	0	0	1	3000
Pershing Co.	Rye Patch	1	0	1	0	0	0	1853
Idaho								
Washington Co.	Crane Creek	1	0	1	0	0	0	8001
Owyhee Co.	Castle Creek	1	0	1	0	0	0	9340
Cassia Co.	Raft River	1	0	0	0	1	0	2840
Utah								
Beaver Co.	Cove Fort	1	0	1	0	0	0	7735
	Roosevelt Hot Springs	1	0	1	0	0	0	6107
	Thermo Hot Springs	1	0	1	0	0	0	7288
Iron Co.	Lund	1	0	1	0	0	0	5857
New Mexico								
Sandoval Co.	Los Alamos	1	1	0	0	0	0	1735
Total:		52	25	17	5	3	2	374,129

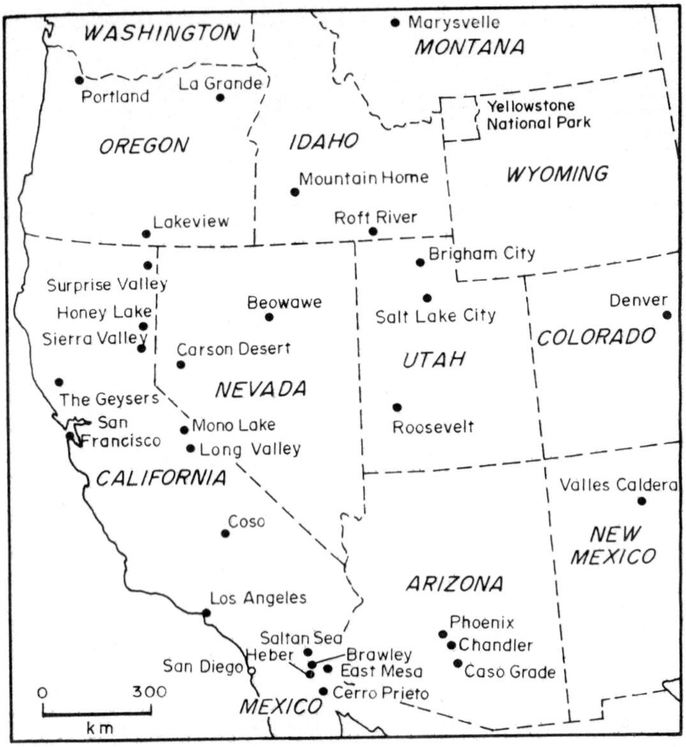

Fig. 8-8. Locations of geothermal drilling in the United States and northern Mexico (from Muffler, 1976b).

Italy. Ceron et al. (1976) have covered in detail the status of geothermal development in Italy, which has the longest history of significant geothermal production. By the end of 1930, the total length of drilling for steam wells had reached 13 km. From 1931 through 1975 an additional 420 km were drilled, and since 1976 the drilling has progressed at a faster rate (Kelly, 1977). Fig. 8-9, from Ceron et al. (1976), shows the geothermal regions of Italy. With a total installed capacity of 418 MW in 1975, Italy is second, after the United States of America, in electric power generation from geothermal energy (Table 8-5). The largest installed electric power generation capacity is at Larderello, where between 1969 and 1975, twenty new productive wells were drilled. However, production from the old wells decreased considerably, causing a net 9% decrease in the steam production. This was offset, in part, by replacing atmospheric turbines by more efficient condensing turbines.

Ente Nazionale per l'Energia Elettrica (ENEL), in cooperation with Consiglio Nazionale di Ricerche (CNR), has carried out extensive exploration throughout the pre-Apennine belt. In the Monte Sabatini region, a well drilled at Cesano in January 1975 encountered temperatures in excess of

210°C. These may exceed 300°C. A hot water geothermal system at a production temperature of 120—140°C has been discovered at Torre Alfina in the Monte Volsini region (Ceron et al., 1976). According to an estimate (Barelli et al., 1975), the electrical energy potential of the pre-Apennine belt is 130—660 MW-centuries. The non-electric use of geothermal resources in Italy is rather limited (Table 8-6).

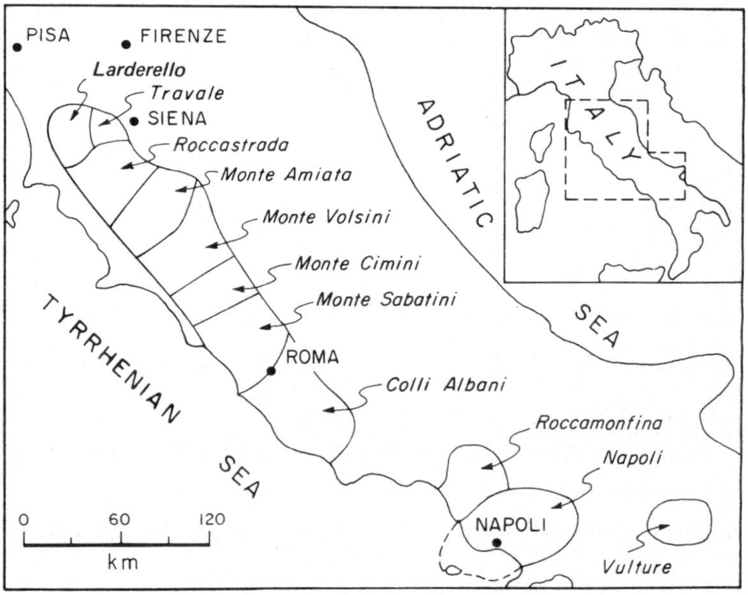

Fig. 8-9. Geothermal regions of Italy (after Ceron et al., 1976).

New Zealand. With a total installed capacity of over 200 MW in 1975, New Zealand is the third largest producer of electrical power from geothermal energy. Approximately one-twelfth of the total electrical energy requirement in New Zealand comes from geothermal resources. Wairakei and Kawerau are the only two producing areas (Fig. 8-10); the installed capacity of Wairakei is 190 MW. Bolton (1976) has described a multiflash system which has increased electrical production capacity at Wairakei by 20 MW. Geothermal fluids are used extensively for domestic heating purpose at Rotorua (Shannon, 1976), and for the air conditioning of a hotel (Reynolds, 1970). The use has successfully increased since 1935, when the first bore was sunk. Now there are more than 700 registered bores in the city, ranging from 15 to 370 m in depth and from 2 to 6 inches in diameter. The temperature varies from 49 to 177°C.

Broadlands is another promising field. Geothermal drilling was suspended in New Zealand during 1971—1973, but in 1973 a program of drilling four wells per year was reactivated. Twenty-eight wells were drilled by 1975 and it was estimated that a 165-MW plant could be kept supplied with the necessary geothermal fluids. The New Zealand Power Planning Committee has recommended a 150-MW plant to be commissioned by 1981 (Bolton, 1976). According to some estimates (Bolton, 1976), 2000 MW of electricity could be produced from the New Zealand geothermal fields.

Fig. 8-10. Major developed geothermal areas of New Zealand (after Muffler, 1976b).

Mexico. In Mexico, geothermal electric power production is primarily limited to the Cerro Prieto geothermal field (Chapter 7). In recent years, extensive investigations, including exploratory drilling, have been carried out at Ixtlan de los Hervores and Los Negritos (Fig. 8-11). Intensive reconnaissance and geological surveys have also been carried out at Los Azufres, La Primavera, and Los Marcos. According to Alonso (1976), the estimated geothermal potential of Mexico is roughly 4000 MW.

Japan. Japan has considerable geothermal resources which, while consisting mainly of wet steam, include an important dry field at Matsukawa. Since 1970, geothermal exploration and development in Japan has experienced a rapid acceleration. The dry steam system at Matsukawa and the hot water system at Otake produce 22 MW and 13 MW, respectively (Fig. 8-12). There are plans to expand production at Matsukawa to 90 MW (Mori, 1976). Recently a 25-MW installation has been put into operation at the Onikobe Caldera (Yamada, 1976) and another 10-MW installation began operating at the Onuma hot water system in 1973 (United Nations, Center for Natural Resources, Energy and Transport, 1976). Drilling is in progress and a 50-MW plant will soon be in operation at Takinoue (Mori, 1976). Another 50-MW

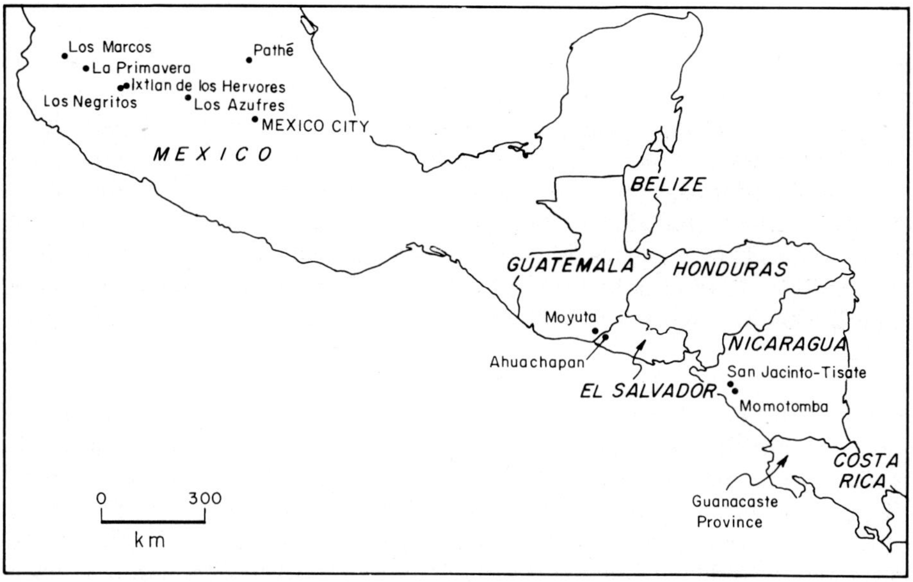

Fig. 8-11. Geothermal areas being explored and developed in Central America (after Muffler, 1976b).

Fig. 8-12. Major geothermal areas of Japan (from Muffler, 1976b).

geothermal power plant is under construction at Hatchobaru (Yamasaki and Hayashi, 1976; Aikawa and Soda, 1976). The islands of Honshu and Hokkaido are under active geothermal exploration. If these ventures succeed, it is expected that by 1982 geothermal electrical capacity at Japan would exceed 1000 MW. In accordance with an agressive program of geothermal exploration undertaken by the Government of Japan, that is a part of the Sunshine Project (Sakakura, 1976), a manyfold increase in geothermal electric generation capacity is expected to be achieved by the year 2000 (Mori, 1976). Geothermal energy also finds use in recreational, agricultural, animal husbandry and space heating activities in Japan (Table 8-6; Howard, 1975).

Iceland. Geothermal waters have been extensively used at Reykjavik and small communities surrounding the capital city for a long time (Fig. 8-13). The Reykjavik heating system is the largest energy facility in the country.

Krafla geothermal field, which was expected to produce 55 MW of electricity by 1977 (Palmason, 1976) continued to be troubled by earthquakes and subsidence (Birsic, 1978). Early in January 1978, a new series of tremors and surface shifting occurred. At certain localities, the ground rose and sank by almost half a meter. While the Krafla plant is ready to generate electricity, adequate steam is not forthcoming from the existing wells. This is attributed to subsidence which has upset the drilling program and clogged many existing wells.

With its location on the Mid-Atlantic Ridge, Iceland has a large geothermal potential, both for electricity generation and space heating. Bodvarsson

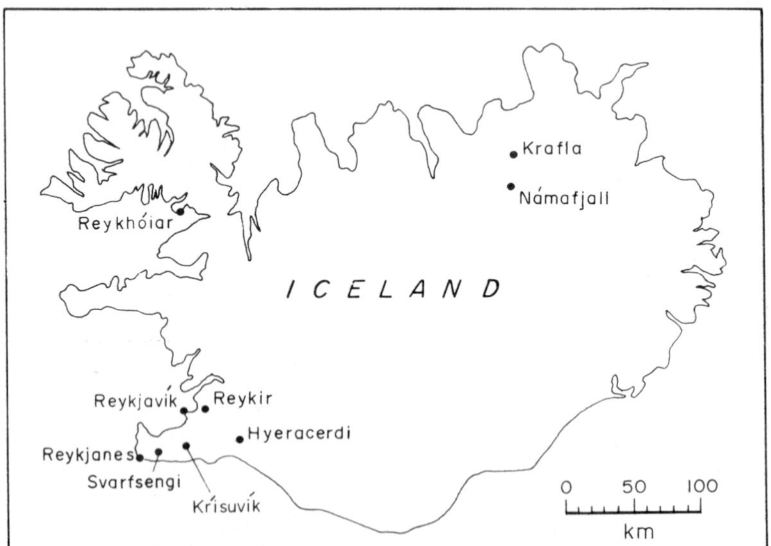

Fig. 8-13. Major developed geothermal areas of Iceland (from Muffler, 1976b).

(1976) has estimated a production potential of 3200 MW for 50 years and the heat content of the recoverable low-temperature resources is equivalent to 4×10^9 metric tons of petroleum.

Latin America. Currently, there exists 135 MW of installed electric generation capacity in Latin America. This includes 75 MW in Mexico and 60 MW in El Salvador. According to Elizarraras (1978) there are two more projects, those at Tatio in Chile and Momotombo in Nicaragua, with definite plans for installing generators in the near future. Another seventeen projects are in the feasibility study stage and six in the reconnaissance stage of study. All this information is displayed in Fig. 8-14 (from Elizarraras, 1978).

Fig. 8-14. Geothermal development in Latin America (from Elizarraras, 1978).

East Africa. Geothermal exploration has been carried out in Afar, primarily being confined to the Assal Rift (Fig. 8-15). The margins of the rift, where any geothermal system will be sealed, are found to be most favorable for geothermal wells (Stieltjes, 1976). Two wells were drilled in 1975, and one of them encountered a temperature of 253°C at 1050 m depth.

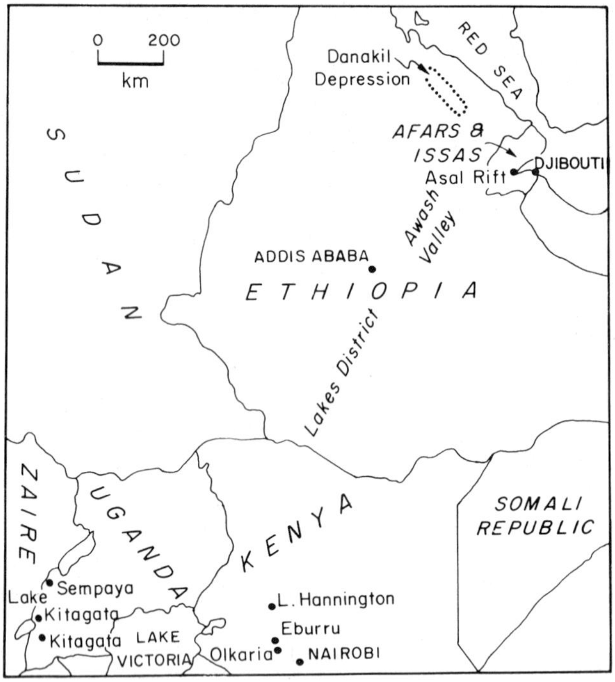

Fig. 8-15. Geothermal areas in East Africa (from Muffler, 1976b).

Under the United Nations Technical Assistance Program, geothermal exploration has been carried out in Ethiopia since 1970. Promising areas have been identified in the Lakes District, the Awash Valley, and the northern Danakil Depression. Work is under way in the Lakes District with a possibility of installing a 10-MW electric power station.

The Uganda Geological Survey and Mines Department has carried out geothermal exploration in the western rift of Uganda. A few resistivity and microearthquake surveys have been carried out at Sempaya, Kitagata, and Lake Kitagata (Fig. 8-15). Maasha (1976) has estimated a subsurface temperature of at least 160°C at Sempaya, the most promising of the three areas.

The rift valley of Kenya is full of hot springs. However, geothermal exploration has only been carried out at Olkaria, Eburru, and Lake Hannington (Fig. 8-15). In addition to the two wells drilled in 1957 and 1958 at Olkaria (Noble and Ojiambo, 1976), four addition holes to depths of 1.3 km have been drilled and temperatures of up to 287°C have been encountered. Further deepening of wells to 1.7 km at Olkaria and fresh drilling at the Lake Hannington and Eburru areas is planned. In view of the high cost of power in Kenya, wells with moderate production are economically attractive (Noble and Ojiambo, 1976).

France. During the middle 1960's, the first large-scale geothermal heating scheme was started in the southeast of Paris, France. The scheme was initiated for 3000 housing units, using the geothermal fluid produced by one 1.8-km-deep borehole (Ten-Dam, 1978). During the period 1975 through 1977, several more sophisticated schemes were initiated. At present, more than 10,000 housing units are heated from geothermal fluids (Ten-Dam, 1978). Jurassic rocks at a depth of 1.8 km in the Paris Basin are the source. Aquifers with temperatures in excess of 50°C have been identified in other sedimentary basins of France (Fouillac et al., 1976).

Turkey. Extensive geothermal exploration surveys have been carried out in western Turkey during the last seven years by the Mineral Research Exploration Institute of Turkey (MTA). It is expected that by the year 2000, geothermal energy would provide 10% of the Turkish electric energy requirement. Up to 1978, sixteen geothermal wells had been drilled at the Kizildere geothermal field (Fig. 8-16). The depths ranged from 370 to 1241 m. The geothermal fluid produced is utilized in operating a 500-kW pilot steam generator and serving a 3000-m^2 greenhouse area. A 15-MW power station is under planning.

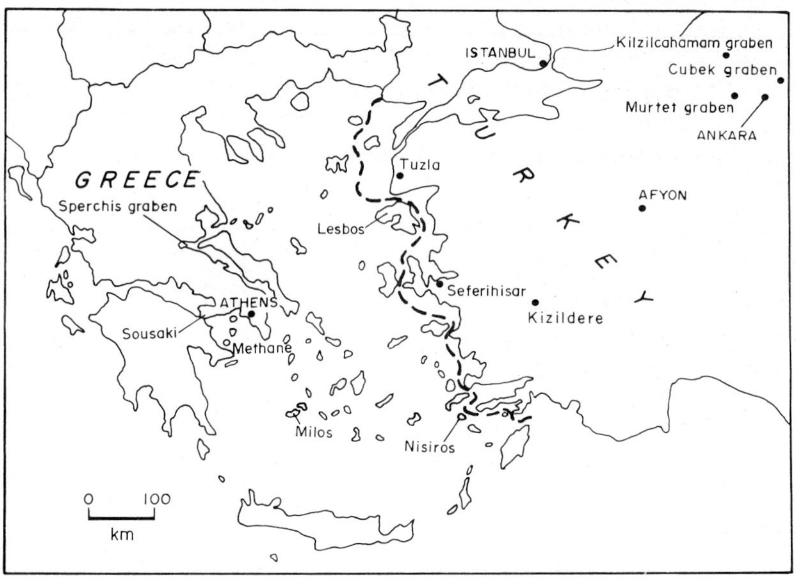

Fig. 8-16. Geothermal exploration in Turkey and Greece (from Muffler, 1976b).

Exploration and drilling are under way in the vicinity of Ankara and Afyon, primarily for space heating purposes. Other interesting areas have been discovered in the Seferihisar and Tuzla regions of western Turkey. A reservoir temperature in excess of 200°C has been inferred in the Seferihisar area (Kurtman and Samilgil, 1976). At Tuzla, a shallow borehole has produced a mixture of steam and water with the bottom hole temperature being 145°C (Öngür, 1975).

Greece. Reconnaissance surveys have been carried out in six areas in Greece (Fig. 8-16). The island of Milos, where a 70-m hole drilled in 1972 discharged steam and water and had a bottom hole temperature of 138°C, appears to be the most promising area (Dominco and Papastamatoki, 1976). Deep test drilling at Milos is under way.

Israel. A detailed five-year plan for geothermal exploration was prepared for Israel in 1976 (Eisenstat, 1978). This plan included various phases of exploration to examine the geothermal potential. Low-temperature (less than 90°C) systems seem common at drillable and reasonably shallow depths, especially adjacent to the Dead Sea Rift. Additional exploration is warranted in the Dead Sea—Jordan Valley—Sea of Galilee area and along Emeq Israel. In the next few years, the full geothermal potential of Israel will be estimated. It appears that multi-purpose use of geothermal energy for non-electric purposes would be more useful in Israel.

Indonesia. During the last several years, extensive geothermal exploration has been carried out in Indonesia, particularly on the islands of Java and Bali (Fig. 8-17). The Kawah-Kamajang area, with a vapor-dominated geothermal system, is found to be very promising (Hochstein, 1976). Several boreholes have been drilled to depths of 500 to 800 m and some of them produce dry steam (Hochstein, 1976; Kartokusumo et al., 1976). A 250-kW electric power station is under installation close to the Kamajang geothermal field.

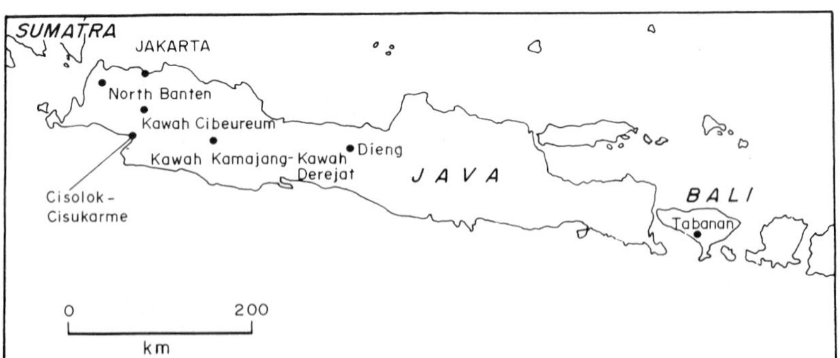

Fig. 8-17. Major geothermal areas under exploration in Indonesia (from Muffler, 1976b).

Unlike the Kawah-Kamajang area, the presence of a hot water field at depth has been inferred for the Dieng area. Exploration is under way in several other areas and according to Radja (1976), substantial geothermal potential has been indicated throughout Indonesia.

Philippines. The island country of the Philippines has made much progress during the last several years in utilizing its vast geothermal energy. According to the report by Birsic (1978), a 110-MW electric power plant was to start functioning by early 1979 in the Tiwi area of southeastern Luzon (Fig. 8-18). Similarly, the Los Banos area is being developed rapidly. Several wells have been drilled and the steam-to-water ratio is found to be better at Los Banos

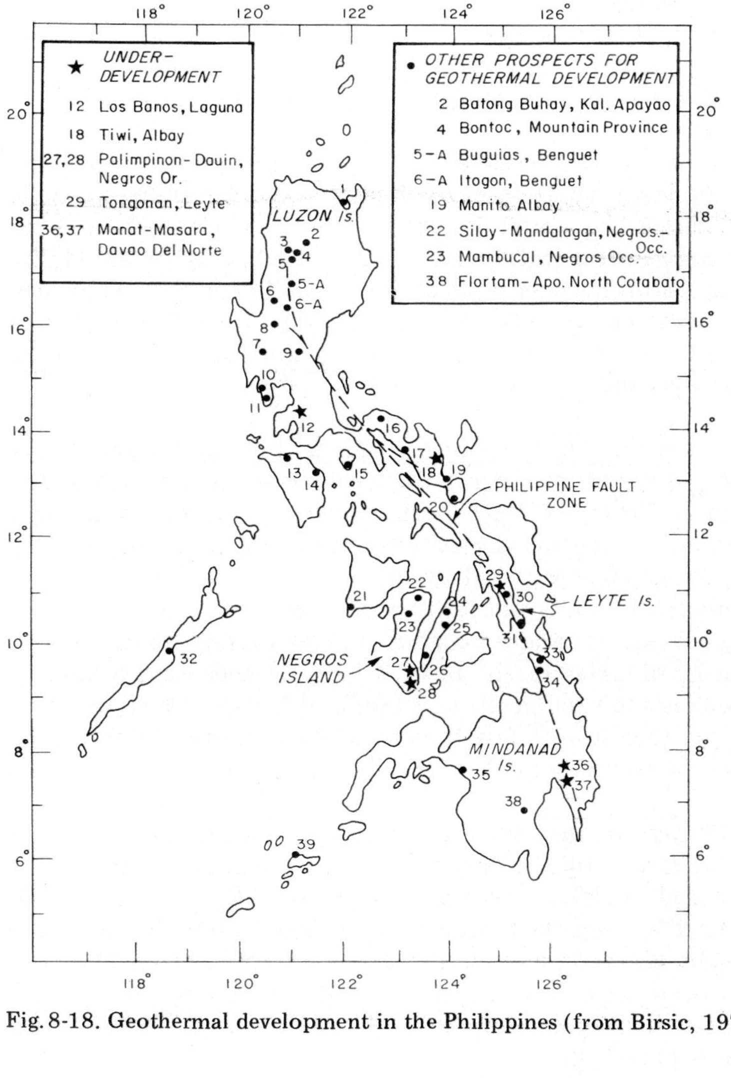

Fig. 8-18. Geothermal development in the Philippines (from Birsic, 1978).

than at Tiwi. The first 110-MW plant is expected to be operative at Los Banos some time during 1979 (Birsic, 1978). The Tongonan geothermal field lies along the Philippine fault zone (Fig. 8-18), extending from Luzon in the north to Mindanao in the south. Drilling carried out during 1976 confirmed the commercial potential of this field and a 3-MW pilot power plant was formally put into operation on July 21, 1977. Further drilling is in progress and it is expected that during 1980 sufficient geothermal fluids would be available to warrant the operation of a 55-MW power plant.

Another important geothermal field in the Philippines is the Southern Negros area. Geological, geophysical and geochemical investigations have indicated the existence of underground fluids at sufficiently high temperatures. The thermal manifestations are essentially controlled by geologic structures. Plans for drilling are under way.

The Davao geothermal field is located in a mining region which is geologically related to the diorite stocks and young volcanic plugs. Resistivity surveys, geothermal gradient measurements, etc., have been carried out and drilling is in progress.

In addition to the above-mentioned five geothermal fields, there are several other equally potential areas in the Philippines. Twenty-five volcanic centers, scattered along the entire length of the archipelago, have been identified by the Sectoral Committee on Geothermal Energy of the Philippines. The existing 46 production wells at Tiwi, Los Banos and Tongonan geothermal fields have an aggregate power capability of 280 MW (Birsic, 1978). The Philippines expects to be one of the major users of geothermal energy in the world by 1985.

Taiwan. There are over sixty known hot spring localities distributed all over Taiwan. Most of the hot springs occur in the volcanic areas, and a few in the metamorphic terrain (MRSO, 1977). Fumaroles occur at a few places. Geophysical, geological and geochemical methods have been applied extensively in the search for geothermal energy in Taiwan.

The Tatun geothermal area is one of the ones with the most potential (Fig. 8-19). Strong geothermal manifestations such as roaring fumaroles and boiling springs can be observed in this area. Drilling has been carried out and the highest recorded reservoir temperature is 300°C at Matsao. The completed exploratory wells produce from a few tons to 33 tons of thermal fluids per hour. The electrical power potential of the Tatun area is estimated to be 100—500 MW.

Tuchang and Chingshui are two other important geothermal areas in Taiwan. They are underlain by the lower Miocene Lushan Formation which is essentially composed of slates intercalated with some thin quartzite beds. Seven test holes and a deep well have been drilled in the Tuchang area (MRSO, 1977). Eight test holes and three deep wells have been drilled in the Chingshui area. These deep wells discharge more than 100 tons of hot water (with 15% steam) per hour. Fluid temperature is 200°C. A pilot plant with 1.5 MW capacity is being set up.

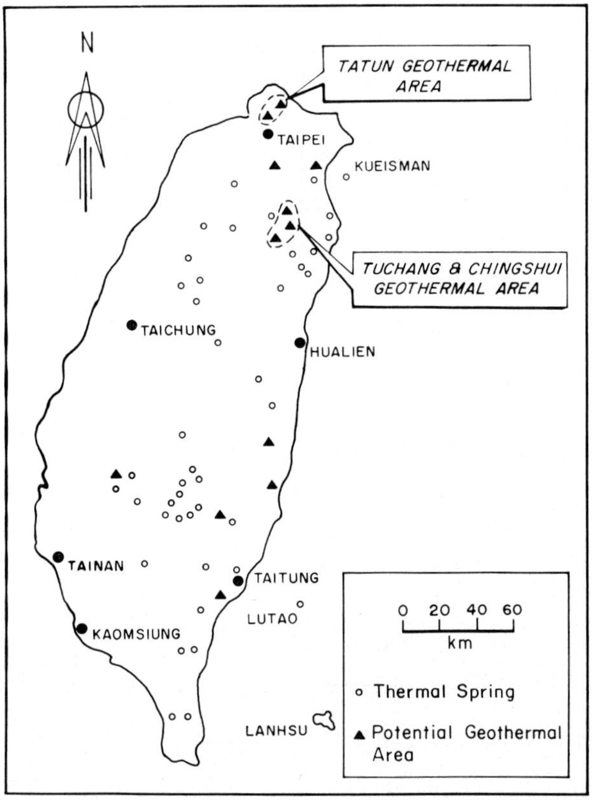

Fig. 8-19. Geothermal areas of Taiwan (from MRSO, 1977).

Geothermal energy is also used in Taiwan for greenhouses, lumber-drying kilns, general laboratories, etc.

Eastern Europe and the U.S.S.R. Fig. 8-20, from Muffler (1976b), shows the geothermal areas of eastern Europe. Estimation of subsurface temperatures carried out by Čermák et al. (1976) shows that temperatures in excess of 40°C exist at 1 km depth over most of southeastern Europe. Temperatures in excess of 50°C exist at a depth of 1 km in the Pannonian Basin of Hungary and the basins north and south of the Caucasus. Geothermal fluid is being used for space heating and agricultural purposes in all these three areas.

Most of the geothermal production in Hungary is from thermal waters contained in highly permeable upper Cenozoic sandstones at depths of up to 2.5 km and with temperatures of up to 150°C (Boldizsár and Korim, 1976). Southeastern Hungary has the maximum utilization of geothermal energy, produced from a belt extending northeast from Szeged to Debrecen (Fig. 8-20). At present, there are over 6000 housing units heated by geothermal

Fig. 8-20. Geothermal regions of eastern Europe (from Muffler, 1976b).

waters in Budapest itself (Stone, 1978). There are many more greenhouses, apartments, hospitals, etc., in eastern Hungary utilizing geothermal energy. At the end of 1974, there were 433 wells in the country producing water at temperatures in excess of 35°C at the wellhead (Boldizsár and Korim, 1976). Balogh (1976) estimates that $5-30 \times 10^{10}$ m^3 of thermal waters could be recovered from depths of 1.5—2.5 km below Hungary.

Geothermal resources, similar to Hungary, also exist in the surrounding countries, i.e., Romania, Czechoslovakia, Austria and Yugoslavia, but there has been little utilization to date (Muffler, 1976b).

Geothermal exploration carried out in the Slovakian Socialist Republic revealed that the central depression of the Danube Basin, southeast of Bratislava, where geothermal fluid at 138°C has been found at depths of 2.5—3.0 km, is the most promising region (Franko and Racičký, 1976).

Water at up to 60°C has been produced from drill holes into granite at depths of 660—750 m at Cieplice in Poland (Dowgiallo, 1976). Although geothermal gradients are not high in Poland, there are adequate opportunities for non-electrical production use of the geothermal resources.

Although specific data is lacking, there are enough indications that utilization of geothermal energy is expanding rapidly in the U.S.S.R. According to Kharahashiyan and Khelkvist (1975), twenty-eight geothermal fields in the

U.S.S.R. are in production, supplying heat for industries, agriculture and houses; and since 1966, more than 200 km of exploratory geothermal wells have been drilled. Steam-water deposits exist in Kamchatka and the Kuril Islands, and hot water fields exist in Kamchatka, the Caucasus, Middle Asia, Kazakhistan and Siberia. According to Fedotov et al. (1976), hydrothermal convection systems with temperatures up to 275°C occur in Kamchatka. These systems have enough discharge of geothermal fluid to support a 500-MW electric power generation station.

Canada. Souther (1976) has given an excellent review of the regional geothermal potential of Canada. Young, silicic volcanic centers in British Columbia, specially Mount Edziza and Meager Mountain, are the areas of greatest potential (Fig. 8-21). A 347-m borehole has been drilled at Meager Mountain and waters at 69°C have been discovered (Nevin and Stauder, 1976). On the basis of chemical analysis, reservoir temperature is estimated to be over 185°C. Further work is in progress.

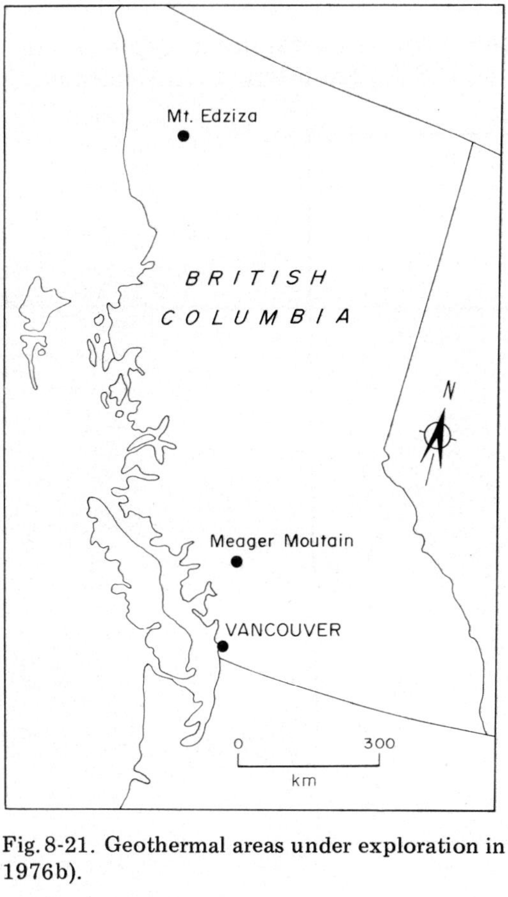

Fig. 8-21. Geothermal areas under exploration in British Columbia, Canada (from Muffler, 1976b).

India. Over most of India, geothermal reconnaissance surveys have been carried out, with special emphasis on the tectonic setting and chemistry of hot springs, to estimate subsurface temperatures (Gupta et al., 1976a; Krishnaswamy, 1976; Subramaniam, 1976). The Himalayan arc in northwestern India, with known geothermal reservoirs at Chumathang, Puga and Parbati Valley, appears to be of the most potential interest (Fig. 8-22). The Konkan area, Cambay, Narbada-Tapti and the Godavari grabens may also have significant potential.

Boreholes drilled in Puga to 80 m depth encountered temperatures of up to 135°C. Chemical analyses of geothermal fluids indicated subsurface temperatures of 220—270°C (Shanker et al., 1976). At Chumathang, a temperature of 109°C was recorded at a depth of 30 m, and a reservoir temperature of 145 to 184°C has been estimated from chemical analyses (Shanker et al., 1976). In Parbati Valley, including the Manikaran area, the geochemistry of fluids suggests reservoir temperatures of over 200°C (Jangi et al., 1976; Gupta et al., 1976b; Chaturvedi and Raymahashay, 1976). It appears that scaling due to deposition of calcium carbonate would pose a major problem in all these three areas.

In the Cambay graben, high pressures and temperatures at depth, conditions similar to the Gulf Coast of the U.S.A., are inferred (Krishnaswamy,

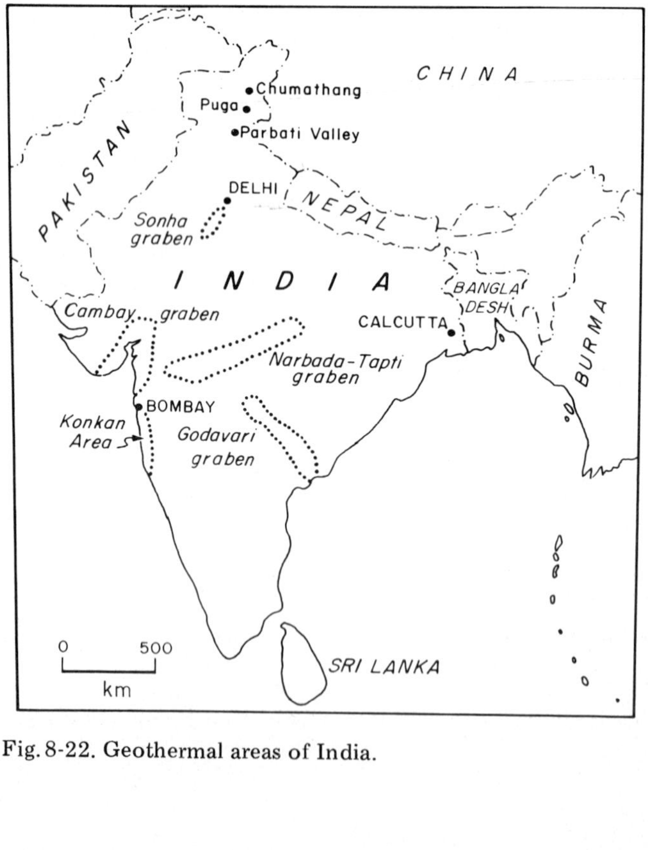

Fig. 8-22. Geothermal areas of India.

1976). A pressure of 100 kg/cm^2 and temperature of 170°C have been encountered at a depth of 3.4 km. It is expected that in the near future, India will make rapid progress in the utilization of geothermal resources.

Egypt. Recently a program of geophysical investigations, including microearthquake activity, geothermal gradients and heat flow, has been undertaken in Egypt (Morgan and Swanberg, 1977). Fig. 8-23 shows the locations of heat flow measurements, thermal manifestations, etc. There does not appear to be a correlation between microearthquake occurrence and geothermal activity. The highest measured geothermal gradient is 54°C/km in the Precambrian granitic rocks at Wadi Ghadir (Fig. 8-23). From the preliminary data it appears that the geothermal resources in Egypt are located on the Red Sea coast and around the Gulf of Suez. Further studies are in progress.

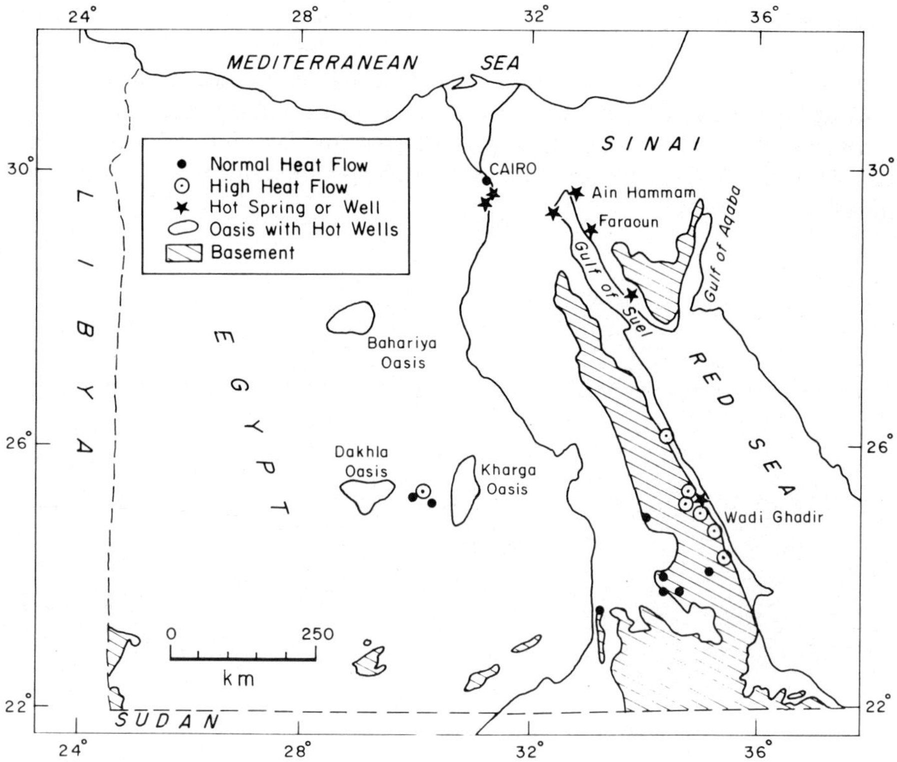

Fig. 8-23. Geothermal studies in Egypt (from Morgan and Swanberg, 1977).

Portugal. Geothermal exploration has been undertaken in the Azores, Portugal. A hole to the depth of 981 m has been drilled on the north flank of the Agua de Pan volcano where temperatures of over 200°C were encountered at a depth of 550 m (Meucke et al., 1974). Further exploration and development are planned.

Iran. A geothermal exploration program is about to begin in the regions of Sabalan, Sahaud, Damavand, Khoy and Maku in northwestern Iran as reported in the *Geothermal Energy Magazine* (1977). The area to be investigated covers approximately 31,000 km^2.

CONCLUDING REMARKS

In the previous sections, available information on the world-wide utilization of geothermal resources at the time of writing has been given. For some countries, it is quite up to date, while for some other countries the necessary information has not been available. As discussed elsewhere, with the passage of time, the use of geothermal energy for electric power generation and for other industrial, agricultural and domestic purposes will increase. According to a conservative estimate, it is expected that by the end of the present century 100,000 MW of electric power would be generated from geothermal resources.

Chapter 9

FURTHER TOPICS

ARTIFICIAL STIMULATION OF GEOTHERMAL SYSTEMS

Extensive drilling carried out all over the world has shown that very often usefully high temperatures are encountered at depth but these drill holes produce no natural steam or hot water. Efforts are being made to extract and utilize the energy content of these hot dry reservoirs economically. One obvious way of accomplishing this objective is to inject water into the hot rock through one hole, allow it to circulate through natural or artificially created flow passages, and recover the steam or hot water through another hole. Major problems encountered are those of avoiding excessive loss of water when the natural permeability is too high, and of creating openings for fluid circulation and enough surface to permit extraction of heat at a reasonable rate for a usefully long duration.

The Los Alamos Scientific Laboratory experiments

A large-scale experimental project has been undertaken by the Los Alamos Scientific Laboratory, Los Alamos, New Mexico, to extract geothermal energy from hot dry rocks at depth (Smith et al., 1976). Fig. 9-1 is a conceptual diagram of the initial Los Alamos concept of extracting geothermal energy by drilling and fracturing. Fig. 9-2 shows the area, west of the Valles Caldera, New Mexico, chosen by the Los Alamos Laboratory for experimentation. Beginning from 1972, a number of holes have been drilled and large surface areas for heat transfer have been created using conventional hydraulic fracturing techniques developed by the oil industry (Pettitt, 1978). On June 3, 1977 during a 20-hour pumping experiment, cold water was pumped down at 1000 psi. The temperature of the recovered water was 130°C, the recovery rate being 85% of the injection rate.

A prolonged circulation and heat extraction experiment was begun in January 1978 to determine the chemical, mechanical and physical properties of the reservoir and the heat exchange system. The circulation was continued for 1800 hours. The results have been favorable. Several aspects of chemical behavior and of flow indicate that thermal contraction and fluid pressurization may have considerably increased the accessible rock volume to fluid circulation (Pettitt, 1978).

As reported by Pettitt (1978), during 1979, the Fenton Hill system will be expanded by drilling to a depth of 3800 m, where temperatures of 250°C are expected. Such a system is expected to be able to produce enough electricity

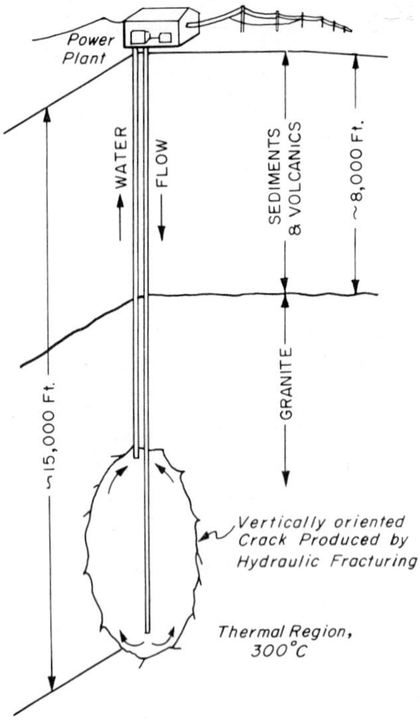

Fig. 9-1. Extraction of geothermal energy from hot dry rock. For details, see text.

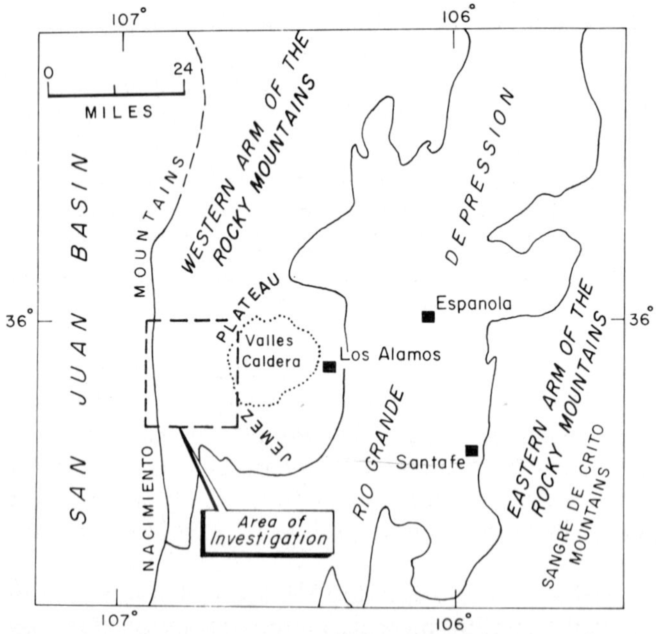

Fig. 9-2. The Valles Caldera, New Mexico: area chosen by the Los Alamos Scientific Laboratory for extracting geothermal energy from hot dry rock. For details, see text.

for 10,000 consumers. The experiment will also provide useful information for the design and construction of small-scale electric generating plants. If the experiment is a success, the first demonstration of utilizing hot rock geothermal energy for generating electricity and other purposes could be in progress by 1985. These systems would be small, each producing 50—100 MW of thermal or 10 MW of electrical power.

Forced geoheat recovery in the Pacific Northwest

Bodvarsson and Reistad (1976) have investigated the economics of forced geoheat recovery in the Cenozoic volcanic areas of the Pacific Northwest, including the Columbia River basin and Snake River plain flood basalt areas. It appears that forced geoheat recovery techniques could successfully be applied and considerable amounts of heat, for a variety of low-temperature uses, may be extracted from the Cenozoic and underlying Mesozoic volcanics. Considering the prevailing economic conditions, forced extraction of geothermal fluids, at temperatures varying from 90 to 150°C from depths of 2.5—3.5 km, appears attractive.

Underground explosions for fracturing

Diadkin and Pariisky (1976) have developed formulas for estimating the radii of the cavity, crushing and fracture zones caused by underground explosions. Actual explosions and laboratory experiments data have been used in developing these formulas. The permeability distribution is also estimated. This analysis shows that underground explosion is a promising stimulating technique.

THE FIRST GEOPRESSURED GEOTHERMAL WELL

The drilling of the first geopressured geothermal well in the Brazoria County, 70 km south of Houston, Texas, began on June 30, 1978. The test well project is expected to cost $8 million over the next two years. The probability of success is estimated to be 75%. The well is a turning point in the four-year-old geothermal research program at the University of Texas at Austin. The Brazoria test well will tap and evaluate the geopressured geothermal energy confined in large bodies of over-pressured hot water trapped in deep sedimentary layers along the Gulf Coast at depths from 7000 to 20,000 feet. The temperatures are predicted to be 325°F and pressures as great as 15,000 psi.

ENVIRONMENTAL FACTORS AND WASTE DISPOSAL

Armstead (1976) has given an excellent summary of environmental factors and waste disposal related with geothermal energy. Initially, geothermal energy enthusiasts widely acclaimed it to be *clean*. There are basically three reasons supporting this view: (1) for a given scale of heat exploitation, geothermal energy is less polluting than fuel combustion; (2) in geothermal areas, even when they are not exploited commercially, there is often natural pollution; and (3) the polluting effects of fuel combustion on a vast and ever-increasing scale have been tolerated for generations. However, in recent years, public concern and awareness of the hazards of all forms of environmental pollution has been aroused and it is necessary to examine and control environmental pollution caused by harnessing geothermal energy.

Hydrogen sulphide

Geothermal fluids almost invariably contain some amount of hydrogen sulphide. In moderate and harmless concentrations, H_2S has an unpleasant smell. When strongly concentrated, it paralyses the olfactory nerves and thus becomes odorless and dangerous. It gives no warning of its presence when it is present in lethal quantities. Being heavier than air, it tends to collect in low-lying pockets. Fatalities are known to have occurred in the vicinity of fumaroles. However, no such case has so far been associated with geothermal exploration or power production. Hydrogen sulphide also has adverse effect on crops and attacks equipment like cummutators, electrical contacts, etc. As listed by Armstead (1976), H_2S escapes to the environment as follows: (1) from condenser gas ejector discharge; (2) with air and warm vapor rising from the cooling tower; (3) well discharge into the atmosphere during testing, or when a plant has surplus inputs of geothermal fluids; (4) from wild bores; (5) from drains and traps; (6) from cooling towers; (7) in solution in cooling water where river cooling is adopted for turbine condensers; and (8) in solution with geothermal water from a wet field when the water is discharged into streams or rivers.

According to the estimates of Axtmann (1976), H_2S discharged daily from the Cerro Prieto 75-MW power station is about 55 tons, and it is expected to be 30 tons for Broadlands, New Zealand, when the 200-MW power station becomes operative. Reed and Campbell (1976) estimate that 28 tons/day of H_2S is discharged at the 500-MW installation at The Geysers, California. Allen and McCluer (1976) have described methods used at The Geysers to tackle escape paths (1), (2), (5) and (6) above. The ejector gases are burnt to convert H_2S into SO_2, which is then scrubbed by the cooling tower water. Elemental sulphur is precipitated as a result of the *Claus reaction*. At the same time a nickle or iron salt which is added to the cooling water acts as a catalyst and thereby precipitates sulphur by oxidizing the H_2S. A certain amount of H_2S is naturally oxidized in the cooling towers. The elemental sulphur is filtered out from the cooling water as a sludge and the surplus cooling water is reinjected. Traps and drains are also piped to the cooling towers where they receive the same treatment as the condensate and cooling water. These methods are quite effective.

The only way to control escape path (3) is to throttle back the wells when a plant is shut down. Escape path (4) could only be controlled by more careful drilling so that wild bores could be avoided as far as possible. No simple solution can be found for escape path (7), except that where river flows are not copious and are far away from the sea, cooling towers should be used instead of direct river cooling.

Carbon dioxide

Carbon dioxide is the main incondensable gas accompanying the geothermal fluids. It escapes into the environment by the same eight abovementioned paths. The emission of large quantities of CO_2 from geothermal installations is inevitable. The fact that fuel combustion produces far greater quantities of carbon dioxide has been an excuse for inaction. With the rapid development of geothermal energy, the problem will gain urgency and will need to be tackled in the near future.

Airborne poisons

In addition to H_2S, harmful elements such as mercury, arsenic compounds and radioactive elements escape into the environment from geothermal installations. However, the quantities escaping into the atmosphere have so far been negligible. Nevertheless, it is advisable to keep a watch and carry out systematic monitoring for these pollutants.

Waterborne poisons

Poisonous elements such as boron, mercury, ammonia and arsenic are often present in geothermal waters. When this water is discharged into rivers or streams, the contaminated downstream waters could be harmful to farming, fisheries or drinking purposes. This problem has been discussed by Axtmann (1976), Anderson (1976), and Rothbaum and Anderton (1976). The possible remedies are reinjection, disposal into seawater where possible, use of evaporator ponds, and storing the water in dry seasons with subsequent release to rivers when they are in spate during the rainy season. It has been also proposed to remove the arsenic, simultaneously with the silica by preoxidizing.

Noise

When steam escapes at high pressures, it generates noise which can be distressing to the ears. It is a common practice for workers to wear ear plugs or muffs when they work close to the new wellhead sites. In a geothermal field under production, noise is generally produced by one or more of the following three mechanisms: (1) due to release of pressurized hot water in wet fields to waste and flashing in the process; (2) due to small quantities of steam being vented to waste to maintain pressure and flow; and (3) due to large quantities of steam being vented when a plant is shut down either because of some trouble or for maintenance. Use of mufflers is found to be an effective way of mitigating these undesirable noises. Reed and Campbell (1976), Swanberg (1976) and Anderson (1976) have carried out comparative studies of noise levels. Legislative action against harmful noise has been taken in many countries.

Heat

Practically all the cooling methods in vogue for geothermal plants release waste heat into the atmosphere. Because of the inherent inefficiency of the geothermal plants, large amounts of heat and moisture could be added from the wet cooling towers to the atmosphere. The total heat added, even from large geothermal power plants, is only a very small fraction of that from natural processes. However, the exact meteorological impact of this small addition is not yet well known. This impact would greatly depend upon the locally prevailing meteorological conditions.

Silica

Silica is one of the most troublesome products of wet geothermal fields. It often occurs in saturated solution with geothermal water at depth. Consequent to reduction of pressure and flashing, silica either precipitates immediately or remains in solution for a limited time. A considerable amount of money is spent annually at the Wairakei geothermal field in cleaning the silica deposits from open-bore water discharge channels from the field to the river. As described by Thórhallsson et al. (1976), in district heating installations, where bore water remains contained for a long time within the pipes, silica scaling becomes a serious problem. The fear of subterranean silica precipitation has often acted as a deterrent in reinjection (Cuellar, 1976).

For removal of silica, Axtmann (1976) has suggested passing super-saturated silica solutions through a sand-filled, fluidized bed heat exchanger where fall in temperature, in combination with sand nucleation centers, should precipitate and remove the silica. Rothbaum and Anderton (1976) have described a pilot plant for treating super-saturated silica solution with slaked lime to produce useful calcium silicate.

Seismicity and subsidence

Substantial ground subsidence can be caused by withdrawing large amounts of subterranean water from a wet field. This can cause tilting and stressing of pipelines and other surface structures. Stilwell et al. (1976) have reported vertical movements of up to 4.5 m during the last 10 years at Wairakei. Additionally, horizontal movements of up to 22.5 cm have been reported during the last 8 years. The main bore water channel at Wairakei has been damaged due to subsidence. Fortunately, the power house is located at a safe distance from the area of subsidence. The most obvious remedy to counter subsidence is reinjection. Swanberg (1976) has also suggested down-hole heat exchangers instead of extracting the natural thermal fluids. The suggestion, however, is as yet only theoretical.

Prolonged geothermal exploitation, subsidence, and reinjection could trigger earthquakes in critically stressed areas. Experiments carried out at the Rangely oil field, Colorado, have demonstrated that injection of fluid can change subsurface pore fluid pressure and thereby trigger earthquakes (Raleigh et al., 1972). This problem has been dealt in detail by Gupta and Rastogi (1976). A number of earthquakes and subsidence have occurred at the Krafla geothermal field in Iceland (Birsic, 1978). However, the earthquakes have not been triggered by geothermal activity but were natural. Reinjection of geothermal fluids is going to gain popularity in the future and its possible seismic effects need to be carefully monitored.

REINJECTION

Reinjection of fluids has since long been proposed as a practical solution for overcoming the problems of pollution, waste disposal, etc., in relation with harnessing geothermal energy. Undesirable aspects of reinjection, such as (1) possible interference with the useful heat output of a producing well, (2) excessive power requirements to pump down the fluids, (3) possibility of decreasing underground permeability by chemical deposition, and (4) possible seismic effects, have been discussed for a long time. However, recent experiments with reinjection have been favourable.

Einarsson et al. (1976) have described successful reinjection carried out at El Salvador. Disposing of the fluids at Ahuachapan geothermal field, El Salvador, was a serious environmental problem in view of their high boron content. During the reinjection experiment, 2×10^6 m^3 of fluids were reinjected into a 952-m deep bore. The substrata being very permeable, no pumping was necessary. In this case, no scaling or any other undesirable effect of reinjection was observed. Einarsson et al. (1976) recommend that the reinjection bore should be at least 1.5 km away from the producing wells. Kubota and Aosaki (1976) have described the reinjection of eight million tons of hot water since 1972 at Otake, Japan, through three wells, the rate of reinjection being 400 tons/hour. The reinjection was adapted to avoid thermal and chemical pollution. The distances of the reinjection wells from the nearest production wells ranged from 150 to 800 m. Kubota and Aosaki (1976) report that no fall in output or temperature of the producing wells was observed. No seismic effect or contamination of ground water was noticed either. The station output, which had declined from 11 to 8.7 MW before injection, has since recovered to 10 MW. The only adverse effect observed was the deposition of silica on the walls of the reinjection well.

Armstead (1976), in his summary, has concluded that reinjection close to the producing zone would sooner or later cause a drop in the temperature of the fluid produced. On the other hand, reinjection at a strategic distance from the producing zone could increase the production and life of a geothermal field.

LEGAL AND INSTITUTIONAL ASPECTS

Anderson (1976) has summarized the legal and institutional aspects of geothermal energy development. Although such problems associated with geothermal energy are universal in nature, some countries like New Zealand, the U.S.A. and Iceland are more advanced than others in finding solutions to them.

The New Zealand Geothermal Energy Act is one of the oldest in the world and was passed in the year 1953. In the definition of geothermal resources, high- and low-temperature reservoirs are separated at 70°C (Dench, 1976). Persons desiring to drill deeper than 61 m require a license to do so. With the acceptance of the Geothermal Energy Regulations in 1961, geothermal safety became a reality. New Zealand geothermal legislation has a unique feature, according to which a single city, Rotorua, through the Rotorua City Geothermal Energy Empowering Act of 1967, is empowered to act as the legal authority to control geothermal prospecting and utilization. This empowers the city to act in place of the central government in most matters.

Aidlin (1976) has described the most pressing problems delaying the development of geothermal resources in the United States. It is taking unusually long for the governmental agencies to establish criteria and accept concepts applicable to geothermal resources. This may partly be due to the recent emphasis placed on protection of the environment. Governmental agencies are worried about the imaginary and exaggerated dangers and have put up conditions and restraints on development even before the real nature of the resource has been ascertained. There are some good signs also, however. For example, Imperial County in California has realistically approached geothermal resource development and this approach is now starting to bear fruit. Aidlin (1976) stresses that a way should be found for developers and legislators to discuss new legislation openly.

In Iceland, the first law that dealt with geothermal energy was enacted in 1923 (Torfason, 1976). This was overshadowed by the Right of Ownership and Use of Geothermal Resources Act of 1940. The Act of 1940 was folded into the Energy Act of 1968, which is the basic legal authority for geothermal development in Iceland. According to this act, the waters of the country are privately owned but their use for power development and other purposes is governed by the state. It is not clear at what depths and temperatures ownership is shifted to the Icelandic government. Torfason (1976) observes that this has not caused any material hindrance to geothermal development in Iceland.

In conclusion, for a rapid and purposeful growth of geothermal energy, governmental agencies have to take a realistic view of the various aspects of geothermal energy and generate legislation favorable for a healthy and rapid growth of the resource.

REFERENCES

Aidlin, J.W., 1976. United States law as it affects geothermal development. In: *Proceedings 2nd U.N. Symposium on the Development and Use of Geothermal Resources*. U.S. Government Printing Office, Washington, D.C., 3: 2353—2357.

Aikawa, K. and Soda, M., 1976. Advanced design in Hatchbaru geothermal power station. *Proceedings 2nd U.N. Symposium on the Development and Use of Geothermal Resources*. U.S. Government Printing Office, Washington, D.C., 3: 1881—1888.

Albright, J.N., 1976. A new and more accurate method for the direct measurement of earth temperature gradients in deep boreholes. In: *Proceedings 2nd U.N. Symposium on the Development and Use of Geothermal Resources*. U.S. Government Printing Office, Washington, D.C., 2: 847—851.

Allen, G.W. and McCluer, H.K., 1976. Abatement of hydrogen sulfide emissions from The Geysers geothermal power plant. In: *Proceedings 2nd U.N. Symposium on the Development and Use of Geothermal Resources*. U.S. Government Printing Office, Washington, D.C., 2: 1313—1315.

Alonso, H., 1976. Potencial Geotérmico de la República Mexicana. In: *Proceedings 2nd U.N. Symposium on the Development and Use of Geothermal Resources*. U.S. Government Printing Office, Washington, D.C., 1: 17—20.

Anderson, D.N., 1976. Summary of Section XII: Legal and institutional aspects. In: *Proceedings 2nd U.N. Symposium on the Development and Use of Geothermal Resources*. U.S. Government Printing Office, Washington, D.C., 1: cxxix—cxxxii.

Armstead, H.C.H., 1973. What is geothermal energy? In: H.C.H. Armstead (Editor), *Geothermal Energy*. UNESCO, Paris, pp. 15—18.

Armstead, H.C.H., 1976. Some unusual ways of developing power from a geothermal field. In: *Proceedings 2nd U.N. Symposium on the Development and Use of Geothermal Resources*. U.S. Government Printing Office, Washington, D.C., 3: 1897—1904.

Arnórsson, S., 1974. The composition of thermal fluids in Iceland and geological features related to the thermal activity. In: L. Kristjansson (Editor), *Geodynamics of Iceland and the North Atlantic Area*. D. Reidel, Dordrecht, pp. 307—323.

Austin, A.L., Higgens, G.H. and Howard, J.H., 1973. The total flow concept for recovery of energy from geothermal hot brine deposits, Lawrence, California. *Lawrence Livermore Lab., Tech. Rep.*, VCRL-51366.

Axtmann, R.C., 1976. Chemical aspects of the environmental impact of geothermal power. In: *Proceedings 2nd U.N. Symposium on the Development and Use of Geothermal Resources*. U.S. Government Printing Office, Washington, D.C., 2: 1323—1327.

Baba, K., Takaki, S., Matsuo, G. and Katagiri, K., 1970. A study of the reservoir at the Matsukawa geothermal field. In: *Proceedings U.N. Symposium on the Development and Utilization of Geothermal Resources. Geothermics, Spec. Issue*, 2(2): 1440—1447.

Baksi, A.K. and Watkins, N.D., 1973. Volcanic production rates: comparison of oceanic ridges, islands and the Columbia plateau basalts. *Science*, 180: 493—496.

Balogh, J., 1976. Results achieved in Hungary in the utilization of geothermal energy. In: *Proceedings 2nd U.N. Symposium on the Development and Use of Geothermal Resources*. U.S. Government Printing Office, Washington, D.C., 1: 29—31.

Banwell, C.J., 1970. Geophysical techniques in geothermal exploration. In: *Proceedings U.N. Symposium on the Development and Utilization of Geothermal Resources. Geothermics, Spec. Issue*, 1: 32—56.

Banwell, C.J., 1973. Geophysical methods in geothermal exploration. In: H.C.H. Armstead (Editor), *Geothermal Energy*. UNESCO, Paris, pp. 31—48.

Banwell, C.J. and Gomezvalle, R., 1970. Geothermal exploration in Mexico 1968—1969. In: *Proceedings U.N. Symposium on the Development and Utilization of Geothermal Resources. Geothermics, Spec. Issue*, 2(1): 27—40.

Barelli, A., Calamai, A. and Cataldi, R., 1975. Estimation of geothermal potential of the pre-Apennine belt of central-southern Italy. *Abstr., 2nd U.N. Symposium on the Development and Use of Geothermal Resources, San Francisco, Calif.*, I-3.

Barr, R.C., 1976. Geothermal energy and electrical power generation. In: *Proceedings 2nd U.N. Symposium on the Development and Use of Geothermal Resources*. U.S. Government Printing Office, Washington, D.C., 3: 1937—1941.

Battelle Columbus and Pacific Northwest Laboratories, 1973. *Environmental Considerations in Future Energy Growth, I. Fuel/Energy Systems*. Technical Summaries and Associated Environmental Burdents, for the Office of Research and Development, Environmental Protection Agency, Columbus, Ohio. Battelle Columbus Laboratories.

Battini, F. and Menut, P., 1964. Etude structurale de la zone loccastrade en vue de la recherche de vapeur pour les méthodes géophysiques, gravimétriques, et électrique. In: *Proceedings U.N. Symposium on New Sources of Energy*. UNESCO, Paris, 2: 73—81.

Birsic, B.J., 1978. Geothermal steam path of New Zealand. *Geotherm. Energy Mag.*, 6(11): 29—32.

Bjornsson, S., Arnórssón, S. and Tomasson, J., 1972. Economic evaluation of Reykjanes thermal brine area, Iceland. *Am. Assoc. Pet. Geol. Bull.*, 56: 2380—2391.

Bjornsson, S. and Einarsson, P., 1974. Seismicity of Iceland. In: L. Kristjánsson (Editor), *Geodynamics of Iceland and the North Atlantic Area*. D. Reidel, Dordrecht, 225 pp.

Bodvarsson, G., 1970. Evaluation of geothermal prospects and the objectives of geothermal exploration. *Geoexploration*, 8: 7—17.

Bodvarsson, G., 1976. Estimates of the geothermal resources of Iceland. In: *Proceedings 2nd U.N. Symposium on the Development and Use of Geothermal Resources*. U.S. Government Printing Office, Washington, D.C., 1: 33—35.

Bodvarsson, G. and Reistad, G.M., 1976. Econometric analysis of forced geoheat recovery for low-temperature uses in the Pacific northwest. In: *Proceedings 2nd U.N. Symposium on the Development and Use of Geothermal Resources*. U.S. Government Printing Office, Washington, D.C., 3: 1559—1564.

Boldizsár, T., 1963. Terrestrial heat flow in the natural steam field at Larderello. *Geofis. Pura Appl.*, 56: 115—122.

Boldizsár, T., 1970. Geothermal energy production from porous sediments in Hungary. In: *Proceedings U.N. Symposium on the Development and Utilization of Geothermal Resources. Geothermics, Spec. Issue*, 2(1): 99—109.

Boldizsár, T. and Korim, K., 1976. Hydrogeology of the Pannonian geothermal basin. In: *Proceedings 2nd U.N. Symposium on the Development and Use of Geothermal Resources*. U.S. Government Printing Office, Washington, D.C., 1: 297—303.

Bolton, R.S., 1976. Recent developments and future prospects for geothermal energy in New Zealand. In: *Proceedings 2nd U.N. Symposium on the Development and Use of Geothermal Resources*. U.S. Government Printing Office, Washington, D.C., 1: 37—42.

Brondi, M., Dall'Aglio, M. and Vitrani, F., 1973. Lithium as a pathfinder element in the large scale hydrogeochemical exploration for hydrothermal systems. *Geothermics*, 2(3/4): 142—153.

Bullard, E., 1973. Basic theories. In: H.C.H. Armstead (Editor), *Geothermal Energy*. UNESCO, Paris, pp. 19—29.

Burgassi, P.D., Ceron, P., Ferrara, G.C., Sestini, G. and Toro, B., 1970. Geothermal gradient and heat flow in the Radicofani region (east of Monte Amiata, Italy). In: *Proceedings U.N. Symposium on the Development and Utilization of Geothermal Resources. Geothermics, Spec. Issue*, 2(1): 443—449.

Cagniard, L., 1953. Basic theory of the magnetotelluric method of geophysical prospecting. *Geophysics*, 18: 605.

Čermák, V., Lubimova, E.A., Stegena, L., 1976. Geothermal mapping in central and eastern Europe. In: *Proceedings 2nd U.N. Symposium on the Development and Use of Geothermal Resources.* U.S. Government Printing Office, Washington, D.C., 1: 47—57.

Ceron, P., Prietro Di Mario, P. and Leardini, T., 1976. Progress report on geothermal development in Italy from 1969 to 1974 and future prospects. In: *Proceedings 2nd U.N. Symposium on the Development and Use of Geothermal Resources.* U.S. Government Printing Office, Washington, D.C., 1: 59—66.

Chapman, D.S. and Pollack, H.N., 1975. Global heat flow: a new look. *Earth Planet. Sci. Lett.*, 28: 23—32.

Chaturvedi, L.N., Raymahashay, B.C., 1976. Geological setting and geochemical characteristics of the Parbati valley geothermal field, India. In: *Proceedings 2nd U.N. Symposium on the Development and Use of Geothermal Resources.* U.S. Government Printing Office, Washington, D.C., 1: 329—338.

Cheng, W.T., 1970. Geophysical exploration in the Tatun volcanic region, Taiwan. In: *Proceedings U.N. Symposium on the Development and Utilization of Geothermal Resources. Geothermics., Spec. Issue*, 2(1): 262—274.

Chilingar, G.U., 1975. An optimistic outlook for world potential source of energy. *Energy Resour.*, 2(2): 229—232.

Combs, J., 1976. Summary of Section IV: Geophysical techniques in exploration. In: *Proceedings 2nd U.N. Symposium on the Development and Use of Geothermal Resources.* U.S. Government Printing Office, Washington, D.C., 1: lxxxi—lxxxvi.

Combs, J. and Muffler, L.J.P., 1973. Exploration for geothermal resources. In: P. Kruger and C. Otte (Editors), *Geothermal Energy*. Stanford University Press, Stanford, Calif., pp. 95—128.

Combs, J., Gupta, H.K. and Helsley, C.E., 1976a. Lateral variations of the subsurface velocity, East Mesa geothermal field, Imperial Valley, California. *EOS Trans. Am. Geophys. Union*, 57: 153 (abstract).

Combs, J., Gupta, H.K' and Jarzabek, D.C., 1976b. Seismic travel time delays and attenuation of anomaly associated with the East Mesa geothermal field, Imperial Valley, California. *EOS Trans. Am. Geophys. Union, Spring Annu. Meet. Progr.*, 34 (abstract).

Combs, J. and Rotstein, Y., 1976. Microearthquake studies at the Coso geothermal area, China lake, California. In: *Proceedings 2nd U.N. Symposium on the Development and Use of Geothermal Resources.* U.S. Government Printing Office, Washington, D.C., 2: 909—916.

Compagnie Générale de Géophysique, 1963. *Master Curves for Electrical Sounding.* Published by the European Association of Exploration Geophysicists.

Cortecci, G., 1974. Oxygen isotopic ratios of sulfate ions water pairs as possible geothermometers. *Geothermics*, 3(2): 60—64.

Corwin, R.F. and Hoover, D.B., 1979. The self potential method in geothermal exploration. *Geophysics*, 44: 226—245.

Craig, H., 1963. The isotopic geochemistry of water and carbon in geothermal areas. In: E. Tongiorgi (Editor), *Nuclear Geology in Geothermal Areas.* Consiglio Nazionale delle Ricerche, Laboratorio di Geologia Nucleare, Pisa, pp. 17—53.

Craig, H., 1976. Isotopic temperatures in geothermal systems. In: *Proc. International Atomic Energy Agency Advisory Group Meeting on the Application of Nuclear Techniques to Geothermal Studies, Pisa, 1975. Geothermics, Spec. Issue*, 5: 8—12.

Craig, H., Boato, G. and White, D.E., 1956. The isotopic geochemistry of thermal waters. In: *Nuclear Geology in Geothermal Areas. Natl. Res. Counc. Nucl. Sci. Ser. Rep.*, 19: 29—44.

Craig, H., Clarke, W.B. and Beg, M.A., 1975. Excess ^{3}He in deep water on the East Pacific Rise. *Earth Planet. Sci. Lett.*, 26: 125—132.

Cremling, J., 1973. Geothermal drilling in California. *J. Pet. Technol.*, 25: 1032—1038.
Creutzburg, H., 1964. Untersuchungen über den Wärmestrom der Erde in Westdeutschland. *Kali Steinsalz*, 4: 73—108.
Cuellar, G., 1976. Behavior of silica in geothermal waste waters. In: *Proceedings 2nd U.N. Symposium on the Development and Use of Geothermal Resources*. U.S. Government Printing Office, Washington, D.C., 2: 1343—1347.

Dall'Aglio, M. and Tonani, F., 1973. Hydrogeochemical exploration for sulfide deposits, correlation between sulfate and other constituents. In: M.J. Jones (Editor), *Geochemical Exploration. Proc. International Geochemical Exploration Symposium, 1972*. Institution of Mining and Metallurgy, London, pp. 305—314.
Dalrymple, G.B., Silver, E.A. and Jackson, E.D., 1973. Origin of the Hawaiian islands. *Am. Sci.*, 61: 294—308.
Deguen, J., Lebouteiller, D., Reford, M. and Tiger, B., 1974. Shallow basin mapping with high resolution aeromagnetics. *44th Meet. Soc. Explor. Geophys., Dallas, Texas* (presented paper).
De la Funte, D.M. and Summer, J.S., 1974. An aeromagnetic study of the Colorado river delta area, Baja California, Mexico. In: *Conference on Geothermal Energy*. California Institute of Technology, Pasadena, Calif. (abstract).
Del Grande, N.K., 1976. An advanced airborne infrared method for evaluating geothermal resources. In: *Proceedings 2nd U.N. Symposium on the Development and Use of Geothermal Resources*. U.S. Government Printing Office, Washington, D.C., 2: 947—953.
Dench, N.D., 1976. The law and geothermal development in New Zealand. In: *Proceedings 2nd U.N. Symposium on the Development and Use of Geothermal Resources*. U.S. Government Printing Office, Washington, D.C., 3: 2359—2362.
Dennis, B.R. and Todd, B.E., 1975. Transducer technology for deep geothermal environments. *8th Transducer Workshop, Dayton, Ohio*.
Diadkin, Y.D. and Pariisky, Y.M., 1976. Theoretical and experimental grounds for utilization of dry rock geothermal resources in the mining industry. In: *Proceedings 2nd U.N. Symposium on the Development and Use of Geothermal Resources*. U.S. Government Printing Office, Washington, D.C., 3: 1609—1618.
Dickinson, D.J., 1973. Aerial infrared survey of Kawerau, Rotorua and Taupo urban areas — 1972. *Geophys. Div., N.Z. Dep. Sci. Ind. Res., Rep.*, 89: 53 pp.
Dickinson, D.J., 1976. An airborne infrared survey of the Tauhara geothermal field, New Zealand. In: *Proceedings 2nd U.N. Symposium on the Development and Use of Geothermal Resources*. U.S. Government Printing Office, Washington, D.C., 2: 955—961.
Dobrin, M.B., Lawrence, P.L. and Sengbush, R.L., 1954. Surface and near surface waves in the Delaware basin. *Geophysics*, 19: 695—715.
Dominco, E. and Papastamatoki, A., 1976. Characteristics of Greek geothermal waters. In: *Proceedings 2nd U.N. Symposium on the Development and Use of Geothermal Resources*. U.S. Government Printing Office, Washington, D.C., 1: 109—121.
Dominguez, A.B. and Vital, B.F., 1976. Repair and control of geothermal wells at Cerro Prieto, Baja California, Mexico. In: *Proceedings 2nd U.N. Symposium on the Development and Use of Geothermal Resources*. U.S. Government Printing Office, Washington, D.C., 2: 1495—1499.
Donaldson, I.G., 1968. The flow of steam water mixtures through permeable beds: a simple simulation of a natural undisturbed hydrothermal region. *N.Z. J. Sci.*, 11: 3—23.
Dowgiallo, J., 1976. The geothermal resources of southwest Poland. In: *Proceedings 2nd U.N. Symposium on the Development and Use of Geothermal Resources*. U.S. Government Printing Office, Washington, D.C., 1: 123—128.
Duffield, W.A., 1975. Late Cenozoic ring faulting and volcanism in the Coso range area of California. *Geology*, 3: 335—338.

Dupart, A. and Omnes, G., 1976. The cost of geophysical programs in geothermal exploration. In: *Proceedings 2nd U.N. Symposium on the Development and Use of Geothermal Resources.* U.S. Government Printing Office, Washington, D.C., 2: 963—970.

Dziewonski, A., Landisman, M., Bloch, S., Sato, Y. and Asano, S., 1968. Progress report on recent improvements in the analysis of surface wave observation. *J. Phys. Earth (Jpn.)*, 16: 1—26.

Eckel, J.E., Deily, F.H. and Ledgerwood, L.W., 1956. Development and testing of jet pump pellet impact drill bits. *AIME Trans.*, 207: 1—9.

Einarsson, S.S., Vides, A.R. and Cuéllar, G., 1976. Disposal of geothermal waste water by reinjection. In: *Proceedings 2nd U.N. Symposium on the Development and Use of Geothermal Resources.* U.S. Government Printing Office, Washington, D.C., 2: 1349—1363.

Eisenstat, S.M., 1978. Geothermal resource development in Israel. *Geotherm. Energy Mag.*, 6(1): 8—18.

Elizarraras, G.R., 1978. OLADE and geothermal energy in Latin America. *Geotherm. Energy Mag.*, 6(7): 35—58.

Ellis, A.J. and Mahon, W.A.J., 1964. Natural hydrothermal systems and experimental hot-water/rock interactions. *Geochim. Cosmochim. Acta*, 28: 1323—1357.

Energy Alternatives — A Comparative Analysis, 1975. The Science and Public Policy Program, University of Oklahoma, Norman, Okla.

Fancelli, R., Nutli. S. and Noto, P., 1975. Chemical and isotopic study of thermal springs in Siena province, Tuscany, Italy. *Abstr., 2nd U.N. Symposium on the Development and Use of Geothermal Resources, San Francisco, Calif.*, III-23.

Faust, C.R. and Mercer, J.W., 1976. Mathematical modelling of geothermal systems. In: *Proceedings 2nd U.N. Symposium on the Development and Use of Geothermal Resources.* U.S. Government Printing Office, Washington, D.C., 3: 1635—1641.

Fedotov, S.A., Balesta, S.T., Droznin, V.A., Masurenkov, Y.P. and Sugrobov, V.M., 1976. On a possibility of heat utilization of the Avachinsky volcanic chamber. In: *Proceedings 2nd U.N. Symposium on the Development and Use of Geothermal Resources.* U.S. Government Printing Office, Washington, D.C., 1: 363—369.

Ferrara, G.C., Ferrara, G. and Gonfiantini, R., 1963. Carbon isotopic composition of carbon dioxide and methane from steam jets of Tuscany. In: E. Tongiorgi (Editor), *Nuclear Geology in Geothermal Areas.* Consiglio Nazionale delle Ricerche, Laboratorio di Geologia Nucleare, Pisa, pp. 277—284.

Fouillac, C., Cailleaux, P. and Michard, G., 1976. Premières études de sources thermales du Massif central français au point de vue géothermique. In: *Proceedings 2nd U.N. Symposium on the Development and Use of Geothermal Resources.* U.S. Government Printing Office, Washington, D.C., 1: 721—726.

Fournier, R.O. and Truesdell, A.H., 1973. An empirical Na-K-Ca geothermometer for natural waters. *Geochim. Cosmochim. Acta*, 37: 1255—1275.

Fournier, R.O. and Truesdell, A.H., 1974. Geochemical indicators of subsurface temperatures, 2. Estimation of temperature and fraction of hot water mixed with cold water. *U.S. Geol. Surv., J. Res.*, 2(3): 263—269.

Franko, O. and Mucha, I., 1976. Geothermal resources of the central depression of the Danubian basin in Slovakia. In: *Proceedings 2nd U.N. Symposium on the Development and Use of Geothermal Resources.* U.S. Government Printing Office, Washington, D.C., 2: 979—992.

Franko, O. and Račický, M., 1976. Present state of development of geothermal resources in Czechoslovakia. In: *Proceedings 2nd Symposium on the Development and Use of Geothermal Resources.* U.S. Government Printing Office, Washington, D.C., 1: 131—137.

Frischknecht, F.G., 1967. Fields about an oscillating magnetic dipole over a two layer earth and application to ground and air borne electromagnetic surveys. *Q. Colo. Sch. Mines*, 62: 326.

Fujii, Y. and Akeno, T., 1970. Chemical prospecting of steam and hot water in the Matsukawa geothermal area. In: *Proceedings U.N. Symposium on the Development and Utilization of Geothermal Resources. Geothermics, Spec. Issue*, 2(1): 1416—1421.

Garcia, S.D., 1976. Geoelectric study of the Cerro Prieto geothermal area, Baja California, Mexico. In: *Proceedings 2nd U.N. Symposium on the Development and Use of Geothermal Resources*. U.S. Government Printing Office, Washington, D.C., 2: 1009—1011.

Geothermal Energy Magazine, 1977. News Briefs, 5(10): 19 pp.

Glover, R.B., 1972. Chemical characteristics of water and steam discharges in the Rift valley of Kenya. *U.N. Kenya Gov. Geotherm. Explor. Proj. Rep.*, August, 106 pp.

Glover, R.B., 1973. Geothermal investigations in Kenya. *N.Z. Geochem. Group News Lett.*, 4(32): 84 pp.

Gonfiantini, R. and Tongiorgi, E. (Editors), 1976. *Proceedings International Atomic Energy Agency Advisory Group Meeting on the Application of Nuclear Techniques to Geothermal Studies. Geothermics, Spec. Issue.*

Gray, G.P. and Young, F.S., Jr., 1973. 25 years of drilling technology — a review of significant accomplishment. *J. Pet. Technol.*, 25: 1347—1354.

Grossling, B.F., 1972. An appraisal of the prospects of geothermal energy in the United States. In: *U.S. Energy Outlook*. National Petroleum Council, Washington, D.C., pp. 15—26.

Gupta, H.K. and Nyman, D.C., 1977a. Short period surface wave dispersion studies of geothermal areas. *EOS Trans. Am. Geophys. Union*, 58: 541 (abstract).

Gupta, H.K. and Nyman, D.C., 1977b. Short period surface wave dispersion studies in the East Mesa geothermal field, California. *Geotherm. Resour. Counc. Trans.*, 1: 123—126.

Gupta, H.K., Nyman, D.C. and Landisman, M., 1977. Shield like upper mantle structure inferred from long period Rayleigh and Love wave dispersion investigations in the Middle East and South East Asia. *Bull. Seismol. Soc. Am.*, 67: 103—119.

Gupta, H.K. and Rastogi, B.K., 1976. *Dams and Earthquakes*. Elsevier Scientific Publishing Company, Amsterdam, 229 pp.

Gupta, M.L., Saxena, V.K. and Sukhija, B.S., 1976a. An analysis of the hot spring activity of the Manikaran area, Himachal Pradesh, India by geochemical studies and tritium concentration of hot spring waters. In: *Proceedings 2nd U.N. Symposium on the Development and Use of Geothermal Resources*. U.S. Government Printing Office, Washington, D.C., 1: 741—744.

Gupta, M.L., Singh, S.B. and Rao, G.V., 1976b. Studies of direct current resistivity in the Puga geothermal field, Himalayas, India. In: *Proceedings 2nd U.N. Symposium on the Development and Use of Geothermal Resources*. U.S. Government Printing Office, Washington, D.C., 2: 1029—1036.

Hagiwara, Y., 1966. Studies of the thermal state of the Earth. The 20th paper: Mountain formation and thermal conduction. *Bull. Earthq. Res. Inst.*, 44: 1537—1551.

Hamilton, R.M. and Muffler, L.J.P., 1972. Microearthquakes at The Geysers geothermal area, California. *J. Geophys. Res.*, 77: 2081.

Hayakawa, M., 1966. Geophysical study of Matsukawa geothermal area, Iwata Prefecture, Japan. *Bull. Volcanol.*, 27: 499—515.

Healy, J., 1970. Pre-investigation geological appraisal of geothermal fields. In: *Proceedings U.N. Symposium on the Development and Utilization of Geothermal Resources. Geothermics, Spec. Issue*, 2(1): 571—577.

Hickel, W.J., 1973. Geothermal energy. In: *Geothermal Resources Research Conference Final Report*, University of Alaska Press, College, Alaska, 95 pp.
Hill, D.P., 1976. Structure of Long Valley caldera, California from a seismic refraction experiment. *J. Geophys. Res.*, 81: 745—753.
Hill, D.P., Mowinckel, P. and Pake, L.G., 1975. Earthquakes, active faults and geothermal areas in the Imperial Valley, California. *Science*, 188: 1306—1308.
Hiramatsu, Y. and Kokado, J., 1958. Eine Untersuchung über die Kühlung von Gruben durch den Wetterstrom. *Bergbau Arch.*, 19: 16—40.
Hochstein, P., 1976. Geophysical exploration of the Kawah Kamojang geothermal field, West Java. In: *Proceedings 2nd U.N. Symposium on the Development and Use of Geothermal Resources*. U.S. Government Printing Office, Washington, D.C., 2: 1049—1058.
Hochstein, M.P. and Dickinson, D.J., 1970. Infrared remote sensing of thermal ground in the Taupo region, New Zealand. In: *Proceedings U.N. Symposium on the Development and Utilization of Geothermal Resources. Geothermics, Spec. Issue*, 2(1): 420—423.
Hochstein, M.P. and Hunt, T.M., 1970. Seismic, gravity and magnetic studies, Broadlands geothermal field, New Zealand. In: *Proceedings U.N. Symposium on the Development and Utilization of Geothermal Resources. Geothermics, Spec. Issue*, 2(1): 333—346.
Hodder, D.T., 1970. Application of remote sensing to geothermal prospecting. In: *Proceedings U.N. Symposium on the Development and Utilization of Geothermal Resources. Geothermics, Spec. Issue*, 2(1): 368—380.
Hodder, D.T., Martin, R.C., Calamai, A. and Cataldi, R., 1973. Remote sensing of Italian geothermal steam field by infrared scanning. In: *Proceedings 1st Pan-American Symposium on Remote Sensing*.
Hoover, D.B. and Long, C.L., 1976. Audio-magneto telluric methods in reconnaissance geothermal exploration. In: *Proceedings 2nd U.N. Symposium on the Development and Use of Geothermal Resources*. U.S. Government Printing Office, Washington, D.C., 2: 1059—1064.
Howard, J.H. (Editor), 1975. Present status and future prospects for nonelectrical uses of geothermal resources. *NATO CCMS Rep.*, No. 40. Lawrence Livermore Laboratory, University of California, Livermore, Calif., 161 pp.
Hulston, J.R. and McCabe, W.J., 1962. Mass spectrometer measurements in the thermal areas of New Zealand, 2. Carbon Isotopic ratios. *Geochim. Cosmochim. Acta*, 26: 399—410.
Hunsbedt, A., London, A.L. and Kruger, P., 1976. Laboratory studies of stimulated geothermal reservoirs. In: *Proceedings 2nd U.N. Symposium on the Development and Use of Geothermal Resources*. U.S. Government Printing Office, Washington, D.C., 3: 1663—1671.

Isachenko, V., Osipova, V. and Sukemel, A., 1969. *Heat Transfer*. Mir Publishers, Moscow, 551 pp.
Iyer, H.M., 1978. Status of seismic methods in geothermal application. In: *Memories from an APSMAGS Workshop*. FT Burgwin Research Center, Taos, N.M., pp. 11—26.
Iyer, H.M., Coppenheimer, D.H. and Hitchcock, T., 1979. Abnormal P-wave delays in The Geysers—Clear Lake geothermal area, California. *Science*, 204: 495—497.
Iyer, H.M. and Hitchcock, T., 1976. Seismic noise survey in Long Valley, California. *J. Geophys. Res.*, 81: 821—840.
Ivanov, V.V., 1967. Principal geochemical environments and processes of the formation of hydrothermal waters in regions of recent volcanic activity. In: A.P. Vinogradov (Editor), *Chemistry of the Earth's Crust*. Israel Program of Scientific Translations, Jerusalem, 2: 260—281.

Jackson, D.D. and Hill, J.H., 1976. Possibilities for controlling heavy metal sulfides in scale from geothermal brines. *UCRL-52977*, Lawrence Livermore Laboratory, University of California, Livermore, Calif., 14 pp.

Jacob, M., 1964. *Heat Transfer*. John Wiley and Sons, New York, N.Y., 1: 758 pp.

Jaegar, J.C., 1971. The effect of the drilling fluid on temperatures measured in boreholes. *J. Geophys. Res.*, 66: 563—569.

James, R., 1976a. Optimum well spacing for geothermal power. In: *Proceedings 2nd U.N. Symposium on the Development and Use of Geothermal Resources*. U.S. Government Printing Office, Washington, D.C., 3: 1681—1683.

James, R., 1976b. Possible serious effect of steam on water flow measurements. In: *Proceedings 2nd U.N. Symposium on the Development and Use of Geothermal Resources*. U.S. Government Printing Office, Washington, D.C., 3: 1703—1706.

James, R., McDowell, G.D. and Allen, M.D., 1970. Flow of steamwater mixtures through a 12-inch diameter pipeline. In: *Proceedings U.N. Symposium on the Development and Utilization of Geothermal Resources. Geothermics, Spec. Issue*, 2: 1581—1597.

Jangi, B.L., Prakash, G., Dua, K.J.S., Thussu, J.L., Dimri, D.B. and Pathak, C.S., 1976. Geothermal exploration of the Parbati valley geothermal field, Kulu district, Himachal Pradesh, India. In: *Proceedings 2nd U.N. Symposium on the Development and Use of Geothermal Resources*. U.S. Government Printing Office, Washington, D.C., 2: 1085—1094.

Jones, P.H., 1970. Geothermal resources of the Northern Gulf of Mexico Basin. In: *Proceedings U.N. Symposium on the Development and Utilization of Geothermal Resources. Geothermics, Spec. Issue*, 2(1): 14—26.

Kappelmeyer, O. and Haenel, R., 1974. *Geothermics with Special Reference to Application*. Gebrüder Borntraeger, Berlin, 238 pp.

Kartokusumo, W., Mahon, W.A.J. and Seal, K.E., 1976. Geochemistry of the Kawah Kamajang geothermal system, Indonesia. In: *Proceedings 2nd U.N. Symposium on the Development and Use of Geothermal Resources*. U.S. Government Printing Office, Washington, D.C., 1: 757—759.

Keller, G.V., 1970. Induction methods in prospecting for hot water. In: *Proceedings Symposium on the Development and Utilization of Geothermal Resources. Geothermics, Spec. Issue*, 2(1): 318—332.

Keller, G.V. and Frischknecht, F.C., 1966. *Electrical Methods in Geophysical Prospecting*. Pergamon, New York, N.Y., 519 pp.

Keller, G.V. and Rapolla, A., 1974. Electric prospecting methods in volcanic areas. In: L. Civetta et al. (Editors), *Physical Volcanology*. Elsevier, Amsterdam, pp. 133—166.

Kelly, A., 1977. A geothermal visit to Italy. *Geotherm. Energy Mag.*, 5(10): 29—32.

Kharahashiyan, F.G. and Khelkvist, V.C., 1975. The present state and the prospects of the utilization of the geothermal resources in the National economy of the USSR. *Abstr., 2nd U.N. Symposium on the Development and Use of Geothermal Resources*, San Francisco, Calif., I-18.

Koeing, J.B., Anderson, D.N. and Hutter, G.W., 1976. Exploration and development of geothermal resources in the United States, 1968—1975. In: *Proceedings 2nd U.N. Symposium on the Development and Use of Geothermal Resources*. U.S. Government Printing Office, Washington, D.C., 1: 139—142.

Koga, A., 1970. Geochemistry of the water discharged from drillholes in the Otake and Hatchobaru areas. In: *Proceedings U.N. Symposium on the Development and Utilization of Geothermal Resources. Geothermics, Spec. Issue*, 2(2): 1422—1425.

Koga, A. and Noda, T., 1976. Geochemical prospecting in vapor-dominated fields for geothermal exploration. In: *Proceedings 2nd U.N. Symposium on the Development and Use of Geothermal Resources*. U.S. Government Printing Office, Washington, D.C., 1: 761—766.

Kononov, V.I. and Polak, B.G., 1976. Indicators of abyssal heat recharge of recent hydrothermal phenomenon. In: *Proceedings 2nd U.N. Symposium on the Development and Use of Geothermal Resources*. U.S. Government Printing Office, Washington, D.C., 1: 767—773.

Krishnaswamy, V.S., 1976. A review of Indian geothermal provinces and their potential for energy utilization. In: *Proceedings 2nd U.N. Symposium on the Development and Use of Geothermal Resources*. U.S. Government Printing Office, Washington, D.C., 1: 143—156.

Kubota, K. and Aosaki, K., 1976. Reinjection of geothermal hot water at the Otake geothermal field. In: *Proceedings 2nd U.N. Symposium on the Development and Use of Geothermal Resources*. U.S. Government Printing Office, Washington, D.C., 2: 1379—1383.

Kurtman, F. and Sâmilgil, E., 1976. Geothermal energy possibilities, their exploration and evaluation in Turkey. In: *Proceedings 2nd U.N. Symposium on the Development and Use of Geothermal Resources*. U.S. Government Printing Office, Washington, D.C., 1: 447—457.

Kusakabe, M., 1974. Sulphur isotopic variations in nature, 10. Oxygen and sulphur isotope study of Wairakei geothermal well discharges. *N.Z. J. Sci.*, 17: 183—191.

Ledgerwood, L.W. Jr., 1960. Efforts to develop improved oil-well drilling methods. *J. Pet. Technol.*, 12: 61—74.

Lewis, C.R. and Rose, S.C., 1970. A theory relating high temperatures and over-pressures. *J. Pet. Technol.*, 22: 11—16.

Lienau, P.J. and Lund, J.W. (Editors), 1974. *Multipurpose Use of Geothermal Energy for Industrial, Agricultural, Commercial and Residential Uses*. Geo-Heat Utilization Center, Oregon Institute of Technology, Oreg., 239 pp.

Lorensen, L.E., Walkup, C.M. and Mones, E.T., 1976. Polymeric and composite materials for use in systems utilizing hot, flowing geothermal brine. In: *Proceedings 2nd U.N. Symposium on the Development and Use of Geothermal Resources*. U.S. Government Printing Office, Washington, D.C., 3: 1725—1731.

Lubimova, E.A., 1967. Theory of thermal state of the earth's mantle. In: T.F. Gaskell (Editor), *Earth's Mantle*. Academic Press, London, pp. 231—323.

Lubimova, E.A., 1969. Thermal history of the earth. In: P.J. Hart (Editor), *The Earth's Crust and Upper Mantle. Am. Geophys. Union, Geophys. Monogr.*, 13: 63—77.

Lyon, G.L., 1974. Geothermal ages. In: I.R. Kaplan (Editor), *Natural Gases in Marine Sediments*. Plenum Press, New York, N.Y., pp. 141—150.

Lyon, G.L. and Hulston, J.R., 1970. Recent carbon isotope and residual gas measurements in relation to geothermal temperatures. *N.Z. Inst. Nucl. Sci. Contrib.*, 407: 11 pp.

Maasha, N., 1976. Electrical resistivity and microearthquake surveys of the Sempaya, Lake Kitagata and Kitagata geothermal anomalies, Western Uganda. In: *Proceedings 2nd U.N. Symposium on the Development and Use of Geothermal Resources*. U.S. Government Printing Office, Washington, D.C., 2: 1103—1112.

Macdonald, W.J.P. and Muffler, L.J.P., 1972. Recent geophysical exploration of the Kawerau geothermal field, North Island, New Zealand. *N.Z. J. Geol. Geophys.*, 15(3): 303.

Mahon, W.A.J., 1973. The chemical composition of natural thermal waters. In: *Proceedings, International Symposium on Hydrogeochemistry and Biogeochemistry, Japan, 1970, 1. Hydrogeochemistry*, J.W. Clark, Washington, D.C., pp. 196—210.

Mahon, W.A.J., 1976. Review of hydrogeochemistry of geothermal systems — prospecting, development and use. In: *Proceedings 2nd U.N. Symposium on the Development and Use of Geothermal Resources*. U.S. Government Printing Office, Washington, D.C., 1: 775—783.

Majer, E.L. and McEvilly, T.V., 1979. Seismological investigations at The Geysers geothermal field. *Geophysics*, 44: 246—269.

Marinelli, G., 1964. Les anomalies thermiques et les champs géothermiques dans le cadre des intrusions récentes en Toscane. In: *Proceedings U.N. Conference on New Resources of Energy*, 2: 288—291.

Marsh, S.E., Lyon, R.J.P. and Honey, F., 1976. Evaluation of NOAA satellite data for geothermal reconnaissance studies. In: *Proceedings 2nd U.N. Symposium on the Development and Use of Geothermal Resources*. UNESCO, Paris, 2: 1135—1141.

Marshall, T. and Braithwaite, W.R., 1973. Corrosion control in geothermal systems. In: H.C.H. Armstead (Editor), *Geothermal Energy*. UNESCO, Paris, pp. 151—160.

Marshall, T. and Tombs, A., 1969. Delayed fracture of geothermal bore casing steels. *Aust. Corros. Eng.*, 13: 2—8.

Matsuo, K., 1973. Drilling for geothermal steam and hot water. In: H.C.H. Armstead (Editor), *Geothermal Energy*. UNESCO, Paris, pp. 73—83.

Maurer, W.C., 1968. *Novel Drilling Techniques*. Pergamon, Oxford.

Maurer, W.C., 1976. Geothermal drilling technology. In: *Proceedings 2nd U.N. Symposium on the Development and Use of Geothermal Resources*. U.S. Government Printing Office, Washington, D.C., 2: 1509—1521.

McDougall, I., 1971. Volcanic island chains and sea floor spreading. *Nature*, 231: 141—144.

McEuen, R.B., 1970. Delineation of geothermal deposits by means of long-spacing resistivity and airborne magnetics. In: *Proceedings U.N. Symposium on the Development and Utilization of Geothermal Resources. Geothermics, Spec. Issue*, 2(1): 295—302.

McEvilly, T., 1978. Problems and controversies of the application of seismic methods to geothermal systems. In: *Memories from an APSMAGS Workshop*. FT Burgwin Research Center, Taos, N.M., pp. 27—33.

McNitt, J.R., 1973. The role of geology and hydrology in geothermal exploration. In: H.C.H. Armstead (Editor), *Geothermal Energy*. UNESCO, Paris, pp. 33—40.

McNitt, J.R., 1976. Summary of United Nations geothermal exploration experiences. In: *Proceedings 2nd U.N. Symposium on the Development and Use of Geothermal Resources*. U.S. Government Printing Office, Washington, D.C., 2: 1127—1134.

Meidav, T., 1970. Application of electrical resistivity and gravimetry in deep geothermal exploration. In: *Proceedings U.N. Symposium on the Development and Utilization of Geothermal Resources, Geothermics, Spec. Issue*, 2(1): 303—310.

Meidav, T., 1972. Electrical resistivity in geothermal exploration. *Annu. Meet. Soc. Explor. Geophys., Anaheim, Calif.* (presented paper).

Meidav, T. and Furgerson, R., 1972. Resistivity studies of the Imperial Valley, California. *Geothermics*, 1(2): 47—62.

Meidav, T. and Rex, R.W., 1970. *Investigation of Geothermal Resources in the Imperial Valley and Their Potential Value for Desalination for Water and Electricity Production*. Institute of Geophysics and Planetary Physics, University of California, Riverside, Calif., 54 pp.

Meidav, T. and Tonani, F., 1976. A critique of geothermal exploration techniques. In: *Proceedings 2nd U.N. Symposium on the Development and Use of Geothermal Resources*. U.S. Government Printing Office, Washington, D.C., 2: 1143—1154.

Mercado, G.S., 1976. Movement of geothermal fluids and temperature distribution in the Cerro Prieto geothermal field, Baja California, Mexico. In: *Proceedings 2nd U.N. Symposium on the Development and Use of Geothermal Resources*. U.S. Government Printing Office, Washington, D.C., 1: 492—498.

Meucke, G.K., Ade-Hall, J.M., Aumento, F., MacDonald, A., Reynolds, P.H., Hyndman, R.D., Quintino, J., Opdyke, N. and Lowrie, W., 1974. Deep drilling in an active geothermal area in the Azores. *Nature*, 252: 281—285.

Mizutani, Y., 1972. Isotopic composition and underground temperatures of the Otake geothermal water, Kyushu, Japan. *Geochem. J.*, 6: 67—73.

Molina, B.R. and Banwell, C.J., 1970. Chemical studies in Mexican geothermal fields. In: *Proceedings U.N. Symposium on the Development and Utilization of Geothermal Resources. Geothermics, Spec. Issue*, 2(2): 1377—1391.

Morgan, W.J., 1972. Plate motions and deep mantle convection. In: R. Shagham (Editor), *Studies in Earth and Space Sciences. Geol. Soc. Am. Mem.*, 132: 7—22.

Morgan, V., 1976. Boron geochemistry. In: R. Thompson (Editor), *Mellor's Comprehensive Treatise of Inorganic Chemistry — Boron Volume (Supplement), Vol. 1.* Longmans-Green, London.

Morgan, P. and Swanberg, C.A., 1977. The Egyptian connection. *Geotherm. Energy Mag.*, 5(10): 7—13.

Mori, Y., 1976. Geothermal resource development program in Northeastern Japan and Hokkaido. In: *Proceedings 2nd U.N. Symposium on the Development and Use of Geothermal Resources.* U.S. Government Printing Office, Washington, D.C., 1: 183—188.

MRSO (Mining Research and Service Organization), 1977. Geothermal energy in Taiwan. *Rep. China, Ind. Tech. Res. Inst., Spec. Publ.*, 12 pp.

Muffler, L.J.P., 1976a. Tectonic and hydrologic control of the nature and distribution of geothermal resources. In: *Proceedings 2nd U.N. Symposium on the Development and Use of Geothermal Resources.* U.S. Government Printing Office, Washington, D.C., 1: 499—507.

Muffler, L.J.P., 1976b. Summary of Section I — present status of resources development. In: *Proceedings 2nd U.N. Symposium on the Development and Use of Geothermal Resources.* U.S. Government Printing Office, Washington, D.C., 1: xxxiii—xliv.

Mundry, E., 1966. Berechnung des gestörten geothermischen Feldes mit Hilfe des Relaxations-Verfahrens. *Z. Geophys.*, 3(32): 157—162.

Nakamura, H., 1969. Mineral and thermal waters of Japan. *23rd Int. Geol. Congr., Prague*, 19: 45—62.

Nathenson, M. and Muffler, L.J.P., 1975. Geothermal resources in hydrothermal convection systems and conduction dominated areas. In: D.E. White and D.L. Williams (Editors), *Assessment of Geothermal Resources of the United States — 1975. U.S. Geol. Surv. Circ.*, 726: 104—121.

National Petroleum Council, 1973. *U.S. Energy Outlook — New Energy Forms.* U.S. Government Printing Office, Washington, D.C., 9 pp.

Nettleton, L.L., 1971. *Elementary Gravity and Magnetics for Geologists and Seismologists.* Society of Exploration Geophysicists, Tulsa, Okla., 121 pp.

Nevin, A.E. and Stauder, J., 1976. Canada — early stages of geothermal investigations in British Columbia. In: *Proceedings 2nd U.N. Symposium on the Development and Use of Geothermal Resources.* U.S. Government Printing Office, Washington, D.C., 2: 1161—1165.

Noble, J.W. and Ojiambo, S.B., 1976. Geothermal exploration in Kenya. In: *Proceedings 2nd U.N. Symposium on the Development and Use of Geothermal Resources.* U.S. Government Printing Office, Washington, D.C., 1: 189—204.

Nusselt, W. and Jurges, W., 1922. Die Kühlung einer ebenen Wand durch einen Luftstrom. *Gesund.-Ing.*, 45: 641—642.

Nyman, D.C. and Landisman, M., 1977. The display equalized filter for frequency-time analysis. *Bull. Seismol. Soc. Am.*, 67: 393—404.

Öngür, T., 1975. Tuzla geothermal field, Cannakkale, Turkey. *Abstr., 2nd U.N. Symposium on the Development and Use of Geothermal Resources, San Francisco, Calif.*, II-36.

Paces, T., 1975. A systematic deviation from Na-K-Ca geothermometer below 75°C and above 10^{-4} atm P_{CO_2}. *Geochim. Cosmochim. Acta*, 39: 541—544.

Palmason, G., 1971. Crustal structure of Iceland from explosion seismology. *Soc. Sci. Islandica Publ.*, 40: 187.

Palmason, G., 1976. Geophysical methods in geothermal exploration. In: *Proceedings 2nd U.N. Symposium on the Development and Use of Geothermal Resources.* U.S. Government Printing Office, Washington, D.C., 2: 1175—1184.

Palmason, G., Friedman, J.D., Williams, R.S. Jr., Jonsson, J. and Saemundsson, K., 1970. Aerial infrared surveys of Reykjanes and Torfajökull thermal areas, Iceland, with a section on coast of exploration surveys. In: *Proceedings U.N. Symposium on the Development and Utilization of Geothermal Resources. Geothermics, Spec. Issue*, 2(1): 399—412.

Panichi, C., Celati, R., Noto, P., Squarci, P., Taffi, L. and Tongiorgi, E., 1974. Oxygen and hydrogen isotope studies of the Larderello (Italy) geothermal system. In: *Isotope Techniques in Ground Water Hydrology*. International Atomic Energy Agency, Vienna, 2: 3—28.

Papadopulos, S.S., Wallace, R.H. Jr., Wesselman, J.B. and Taylor, R.E., 1975. Assessment of onshore geopressured-geothermal resources in the Northern Gulf of Mexico Basin. In: D.E. White and D.L. Williams (Editors), *Assessment of Geothermal Resources of the United States — 1975. U.S. Geol. Surv. Circ.*, 726: 125—146.

Parasnis, D.S., 1970. Some recent geoelectric measurements in the Swedish sulfide ore fields illustrating scope and limitations of the methods concerned. In: *Mining and Ground Water Geophysics*, 1967. Geological Survey of Canada, Ottawa, Ont., pp. 290—301.

Pettitt, R.A., 1978. Hot dry rock: a new potential for energy. *Geotherm. Energy Mag.*, 6(11): 11—19.

Pollastri, G., 1970. Design and construction of steam pipelines. In: *Proceedings U.N. Symposium on the Development and Utilization of Geothermal Resources. Geothermics, Spec. Issue*, 2(1): 780—811.

Radja, V., 1976. Overview of geothermal energy studies in Indonesia. In: *Proceedings 2nd U.N. Symposium on the Development and Use of Geothermal Resources*. U.S. Government Printing Office, Washington, D.C., 1: 233—240.

Ragnars, K., Saemundsson, K., Benediktsson, S. and Einarsson, S.S., 1970. Development of the Namafjal area, Northern Iceland. In: *Proceedings U.N. Symposium on the Development and Utilization of Geothermal Resources. Geothermics, Spec. Issue*, 2(1): 925—935.

Raleigh, C.B., Healy, J.H. and Bredehoeft, H.D., 1972. Faulting and crustal stress at Rangely, Colorado. In: H.C. Heard, I.Y. Bord, N.L. Carter and C.B. Raleigh (Editors), *Flow and Fracture of Rocks. Am. Geophys. Union, Geophys. Monogr.*, 16: 275—284.

Ramey, H.J., Jr., 1976. Pressure transient analysis for geothermal wells. In: *Proceedings 2nd U.N. Symposium on the Development and Use of Geothermal Resources*. U.S. Government Printing Office, Washington, D.C., 3: 1749—1757.

Ramey, H.J., Jr., Kruger, P., London, A.L. and Brigham, W.E., 1976. Geothermal reservoir engineering research at Stanford University. In: *Proceedings 2nd U.N. Symposium on the Development and Use of Geothermal Resources*. U.S. Government Printing Office, Washington, D.C., 3: 1763—1771.

Reed, M.J., 1976. Geology and hydrothermal metamorphism in the Cerro Prieto geothermal field, Mexico. In: *Proceedings, 2nd U.N. Symposium on Development and Use of Geothermal Resources*. U.S. Government Printing Office, Washington, D.C., 1: 539—547.

Reed, M.J. and Campbell, G.E., 1976. Environmental impact of development in The Geysers geothermal field, U.S.A. In: *Proceedings 2nd U.N. Symposium on the Development and Use of Geothermal Resources*. U.S. Government Printing Office, Washington, D.C., 2: 1399—1410.

Renner, J.L., White, D.E. and Williams, D.L., 1975. Hydrothermal convection systems. In: D.E. White and D.L. Williams (Editors), *Assessment of Geothermal Resources of the United States — 1975. U.S. Geol. Surv. Circ.*, 726: 5—57.

Rex, R., 1978. U.S. geothermal industry in 1978. *Geotherm. Energy Mag.*, 6(7): 28—33.

Rex, R.W. and Howell, D.J., 1973. Assessment of U.S. geothermal resources. In: P. Kruger and C. Otte (Editors), *Geothermal Energy*. Stanford University Press, Stanford, Calif., pp. 59—60.

Reynolds, G., 1970. Cooling with geothermal head. In: *Proceedings U.N. Symposium on the Development and Utilization of Geothermal Resources. Geothermics, Spec. Issue*, 2(2): 1658—1661.

Richter, C.F., 1958. *Elementary Seismology*. W.H. Freeman and Co., San Francisco, Calif., 768 pp.

Rightmire, C.T. and Truesdell, A.H., 1974. Carbon isotope composition of soil gases as indicator of geothermal areas. *Geol. Soc. Am. Abstr. Progr.*, 6(7): 927 (abstract).

Risk, G.F., Macdonald, W.J.P. and Dawson, G.B., 1970. D.C. resistivity surveys of the Broadlands geothermal region, New Zealand. In: *Proceedings U.N. Symposium on the Development and Utilization of Geothermal Resources. Geothermics, Spec. Issue*, 2(1): 287—294.

Roberts, A.A., Friedman, I., Donovan, J.J. and Denton, E.H., 1975. Helium surveys, a possible technique for locating geothermal reservoirs. *Geophys. Res. Lett.*, 2: 209—210.

Robinson, B.W., 1973. Sulphur isotope equilibrium during sulphur hydrolysis at high temperatures. *Earth Planet. Sci. Lett.*, 18: 443—450.

Robinson, P.T., Elders, W.A. and Muffler, L.J.P., 1976. Quaternary volcanism in Salton Sea geothermal field. *Geol. Soc. Am. Bull.*, 87: 347—360.

Robinson, R.J. and Morse, R.A., 1976. A study of the effects of various reservoir parameters on the performance of geothermal reservoirs. In: *Proceedings 2nd U.N. Symposium on the Development and Use of Geothermal Resources*. U.S. Government Printing Office, Washington, D.C., 3: 1773—1779.

Rothbaum, H.P. and Anderton, B.H., 1976. Removal of silica and arsenic from geothermal discharge waters by precipitation of useful calcium silicates. In: *Proceedings 2nd U.N. Symposium on the Development and Use of Geothermal Resources*. U.S. Government Printing Office, Washington, D.C., 2: 1417—1425.

Sakakura, S., 1976. Government activity report on geothermal energy in Japan. In: *Proceedings 2nd U.N. Symposium on the Development and Use of Geothermal Resources*. U.S. Government Printing Office, Washington, D.C., 3: 2431—2434.

Schlanger, S.O. and Combs, J., 1975. Hydrocarbon potential of marginal basins bounded by an island arc. *Geology*, 397—400.

Schock, R.N. and Duba, A., 1975. The effect of electrical potential on scale formation in Salton Sea brine. *UCRL-51944*, Lawrence Livermore Laboratory, University of California, Livermore, Calif., 14 pp.

Schoen, R. and Rye, R.O., 1970. Sulfur isotope distribution in solfataras, Yellowstone National Park. *Science*, 170: 1082—1084.

Sestini, G., 1970. Heat flow measurements in non-homogeneous terrains: its application to geothermal areas. In: *Proceedings U.N. Symposium on the Development and Utilization of Geothermal Resources. Geothermics, Spec. Issue*, 2(1): 424—436.

Shanker, R., Padhi, R.N., Arora, C.L., Prakash, G., Thussu, J.L. and Dua, K.J.S., 1976. Geothermal exploration of the Puga and Chumathan geothermal fields, Ladakh, India. In: *Proceedings 2nd U.N. Symposium on the Development and Use of Geothermal Resources*. U.S. Government Printing Office, Washington, D.C., 1: 245—258.

Shannon, R.J., 1976. Geothermal heating of government buildings in Rotorua. In: *Proceedings 2nd U.N. Symposium on the Development and Use of Geothermal Resources*. U.S. Government Printing Office, Washington, D.C., 3: 2165—2172.

Shaw, H.R. and Jackson, E.D., 1973. Linear island chains in the Pacific: results of thermal plumes or gravitational anchors. *J. Geophys. Res.*, 78: 8634—8652.

Sigvaldason, G.E., 1973. Geochemical methods in geothermal exploration. In: H.C.H. Armstead (Editor), *Geothermal Energy*. UNESCO, Paris, pp. 49—59.

Simmons, G., 1967. Interpretation of heat flow anomalies, 1. Contrast in heat production. *Rev. Geophys.*, 5(1): 42—52.

Smith, J.H., 1973. Collection and transmission of geothermal fluids. In: H.C.H. Armstead (Editor), *Geothermal Energy*. UNESCO, Paris, pp. 97—106.

Smith, J.L., Isselhardt, C.F. and Matlick, J.S., 1978. Summary of 1977 geothermal drilling, Western United States. *Geotherm. Energy Mag.*, 6(5): 11—19.

Smith, M.C., Aamodt, R.L., Potter, R.M. and Brown, D.W., 1976. Man-made geothermal reservoirs. In: *Proceedings 2nd U.N. Symposium on the Development and Use of Geothermal Resources*. U.S. Government Printing Office, Washington, D.C., 3: 1781—1787.

Smith, R.L. and Shaw, H.R., 1973. Volcanic rocks as geological guides to geothermal exploration and evaluation. *EOS Trans. Am. Geophys. Union*, 54: 1213.

Smith, R.L. and Shaw, H.R., 1975. Igneous related geothermal systems. In: D.E. White and D.L. Williams (Editors), *Assessment of Geothermal Resources of the United States — 1975. U.S. Geol. Surv. Circ.*, 726: 58—83.

Smith, M., Potter, R., Brown, D. and Aamodt, R.L., 1973. Induction and growth of fractures in hot rock. In: P. Kruger and C. Otte (Editors), *Geothermal Energy*. Stanford University Press, Stanford, Calif., pp. 251—268.

Soda, M., Takahashi, Y., Aikawa, K., Kubota, K. and Eijima, Y., 1976. Experimental study on transmission phenomenon in steam water mixtures flowing through a large pipe line for geothermal power station. In: *Proceedings 2nd U.N. Symposium on the Development and Use of Geothermal Resources*. U.S. Government Printing Office, Washington, D.C., 3: 1789—1795.

Sørensen, B., 1975. Energy and resources. *Science*, 189: 255—260.

Souther, J.G., 1976. Geothermal potential of Western Canada. In: *Proceedings 2nd U.N. Symposium on the Development and Use of Geothermal Resources*. U.S. Government Printing Office, Washington, D.C., 1: 259—267.

Stanley, W.D., Jackson, D.B. and Hearn, B.C. Jr., 1973. Preliminary results of geoelectrical investigations near Clear Lake, California. *U.S. Geol. Surv. Open-File Rep.*, 20 pp.

Stauffer, R.E. and Jenne, E.A., 1975. Arsenic-phosphorus chemistry of Yellowstone geothermal waters. *Abstr., 2nd U.N. Symposium on the Development and Use of Geothermal Resources, San Francisco, Calif.*, IV-14.

Steeples, D.W. and Iyer, H.M., 1976. Teleseismic P-wave delays in geothermal exploration. In: *Proceedings 2nd U.N. Symposium on the Development and Use of Geothermal Resources*. U.S. Government Printing Office, Washington, D.C., 2: 1199—1206.

Stieltjes, L., 1976. Research for a geothermal field in a zone of oceanic spreading: example of the Asal Rift (French Territory of the Afars and the Issas, Afar Depression, East Africa). In: *Proceedings 2nd U.N. Symposium on the Development and Use of Geothermal Resources*. U.S. Government Printing Office, Washington, D.C., 1: 613—623.

Stilwell, W.B., Hall, W.K. and Tawhai, J., 1976. Ground movement in New Zealand geothermal field. In: *Proceedings 2nd U.N. Symposium on the Development and Use of Geothermal Resources*. U.S. Government Printing Office, Washington, D.C., 2: 1427—1434.

Stoces, B. and Cernik, A., 1931. *Bekampfung hoher Grubentemperaturen*. Springer, Berlin.

Stone, A.M., 1978. Geothermal in Hungary. *Geotherm. Energy Mag.*, 6(11): 27—28.

Strangway, D.C., Swift, C.M., Jr. and Holmer, R.C., 1973. The application of audiofrequency magnetotellurics (AMT) to mineral exploration. *Geophysics*, 38(6): 1159—1175.

Studt, F.E., 1964. Geophysical prospecting in New Zealand's hydrothermal fields. In: *Proceedings U.N. Conference on New Sources of Energy*. UNESCO, Paris, 2(1): 380 pp.

Subramanian, S.A., 1976. Present status of geothermal resource development in India. In: *Proceedings 2nd U.N. Symposium on the Development and Use of Geothermal Resources*. U.S. Government Printing Office, Washington, D.C., 1: 269—271.

Sumi, K. and Maeda, K., 1973. Hydrothermal alteration of main productive formation of the steam for power at Matsukawa, Japan. *Proceedings, International Symposium on*

Hydrogeochemistry and Biogeochemistry, Japan, 1970, 1. Hydrogeochemistry. J.W. Clark, Washington, D.C., pp. 211—228.

Swanberg, C.A., 1974. The application of the Na-K-Ca geothermometer to thermal areas of Utah and the Imperial Valley, California. *Geothermics*, 3: 53—59.

Swanberg, C.A., 1976. The Mesa geothermal anomaly, Imperial Valley, California: a comparison and evaluation of results obtained from surface geophysics and deep drilling. In: *Proceedings 2nd U.N. Symposium on the Development and Use of Geothermal Resources*. U.S. Government Printing Office, Washington, D.C., 2: 1217—1229.

Swanson, D.A., 1972. Magma supply rate at Kilauea volcano, 1952—1971. *Science*, 175: 169—170.

Takahashi, Y., Hayashida, T., Soezima, S., Aramaki, S. and Soda, M., 1970. An experiment on pipeline transportation of steam-water mixtures at Otake geothermal field. In: *Proceedings U.N. Symposium on the Development and Utilization of Geothermal Resources. Geothermics, Spec. Issue*, 2(1): 882—891.

Teknekron, Inc., 1973. *Fuel Cycles for Electrical Power Generation Phase I Towards Comprehensive Standards: The Electric Power Case*. Report for the Office of Research and Monitoring, Environmental Protection Agency, Berkeley, Calif.

Ten-Dam, A., 1978. Geothermal heating in France. *Geotherm. Energy Mag.*, 6(4): 11—19.

Thompson, G.A. and Burke, D.B., 1973. Rate and direction of spreading in Dixie Valley, Basin and Range province, Nevada. *Geol. Soc. Am. Bull.*, 84: 627—633.

Thórhallsson, S., Ragnars, K., Arnorsson, S. and Kristmannsdóttir, H., 1976. Rapid scaling of silica in two district heating systems. In: *Proceedings 2nd U.N. Symposium on the Development and Use of Geothermal Resources*. U.S. Government Printing Office, Washington, D.C., 2: 1445—1449.

Tolivia, M.E., 1976. Evaluation of the geothermal potential of Cerro Prieto, Baja California (Mexico). In: *Proceedings 2nd U.N. Symposium on the Development and Use of Geothermal Resources*. U.S. Government Printing Office, Washington, D.C., 1: 279—281.

Tolivia, M.E., Hoashi, J. and Miyazaki, M., 1976. Corrosion of turbine materials in geothermal steam environment in Cerro Prieto, Mexico. In: *Proceedings 2nd U.N. Symposium on the Development and Use of Geothermal Resources*. U.S. Government Printing Office, Washington, D.C., 3: 1815—1820.

Torfason, H., 1976. The law of Iceland as it affects geothermal development. In: *Proceedings 2nd U.N. Symposium on the Development and Use of Geothermal Resources*. U.S. Government Printing Office, Washington, D.C., 3: 2435—2437.

Truesdell, A.H., 1976. Summary of section III: geochemical techniques in exploration. In: *Proceedings 2nd U.N. Symposium on the Development and Use of Geothermal Resources*. U.S. Government Printing Office, Washington, D.C., 1: liii—lxxx.

Truesdell, A.H. and Fournier, R.O., 1976. Deep conditions in the geothermal system of Yellowstone Park, Wyoming from chemical isotopic and geophysical data. *U.S. Geol. Surv. Open-File Rep.*

Uchiyama, M. and Matsuura, S., 1970. Measurement and transmission of steam in Matsukawa geothermal power plant. In: *Proceedings Symposium on the Development and Utilization of Geothermal Resources. Geothermics, Spec. Issue*, 2(2): 1572—1580.

United Nations, Centre for Natural Resources, Energy and Transport, 1976. Present status of world geothermal development. In: *Proceedings 2nd U.N. Symposium on the Development and Use of Geothermal Resources*. U.S. Government Printing Office, Washington, D.C., 1: 3—9.

United Nations, Statistical Office. *Demographic Year Book*. This annual compilation is the source for world data on population.

United Nations, Statistical Office. *World Energy Supplies Series, 1970—1973*.

U.S. Geological Survey, 1975. Assessment of geothermal resources of the United States — 1975. In: D.E. White and D.L. Williams (Editors), *Assessment of Geothermal Resources of the United States — 1975. U.S. Geol. Surv. Circ.*, 726: 155.

Ushijima, K., Nishikawa, K., Ito, T., Matsumoto, K. and Noguchi, T., 1976. A thermometer for geothermal thermometry in geothermal wells. In: *Proceedings 2nd U.N. Symposium on the Development and Use of Geothermal Resources*. U.S. Government Printing Office, Washington, D.C., 3: 1829—1834.

Vakin, E.A., Polak, B.G., Sugrobov, V.M., Erlikh, E.N., Belousov, V.I. and Pilipenko, G.F., 1970. Recent hydrothermal systems of Kamchatka. In: *Proceedings Symposium on the Development and Utilization of Geothermal Resources. Geothermics, Spec. Issue*, 2(2): 1116—1133.

Vanyan, L.L., 1967. *Electromagnetic Depth Soundings*. Consultants Bureau, New York, N.Y., 312 pp.

Wahl, E. and Yen, I., 1976. Scale deposition and control research for geothermal utilization. In: *Proceedings 2nd U.N. Symposium on the Development and Use of Geothermal Resources*. U.S. Government Printing Office, Washington, D.C., 3: 1855—1864.

Ward, P.L., 1972. Microearthquakes: prospecting tool and possible hazard in the development of geothermal resources. *Geothermics*, 1: 3—12.

Ward, P.L. and Bjornsson, S., 1971. Microearthquake swarms and the geothermal areas of Iceland. *J. Geophys. Res.*, 76: 3953—3982.

Ward, P.L., Palmason, G. and Drake, C., 1969. Microearthquake surveys and the Mid-Atlantic Ridge of Iceland. *J. Geophys. Res.*, 74: 665—684.

White, D.E., 1965. Geothermal energy. *U.S. Geol. Surv. Circ.*, 519: 17 pp.

White, D.E., 1968. Environments of generation of some base metal ore deposits. *Econ. Geol.*, 63: 301—335.

White, D.E., 1969. Rapid heat flow surveying of geothermal area, utilizing individual snowfalls as calorimeters. *J. Geophys. Res.*, 74: 5191—5201.

White, D.E., 1970. Geochemistry applied to the discoveries, evaluation, and exploitation of geothermal energy resources. Rapporteurs Report. In: *Proceedings U.N. Symposium on the Development and Utilization of Geothermal Resources. Geothermics. Spec. Issue*, 1: 58—80.

White, D.E., 1973. Characteristics of geothermal resources. In: P. Kruger and C. Otte (Editors), *Geothermal Energy*. Stanford University Press, Stanford, Calif., pp. 68—92.

White, D.E., 1974. Characteristics of geothermal resources. In: P. Kruger and C. Otte (Editors), *Geothermal Energy*. Stanford University Press, Stanford, Calif., pp. 69—94.

White, D.E., Hem, J.D. and Waring, G.A., 1963. Chemical composition of subsurface waters. In: M. Fleischer (Editor), *Data of Geochemistry. U.S. Geol. Surv. Prof. Paper*. 440-F: 1—67.

White, D.E., Muffler, L.P.J. and Truesdell, A.H., 1971. Vapor-dominated hydrothermal systems compared with hot-water systems. *Econ. Geol.*, 66: 75.

White, D.E. and William, D.L., 1975. Summary and conclusions. In: D.E. White and D.L. Williams (Editors), *Assessment of Geothermal Resources of the United States — 1975. U.S. Geol. Surv. Circ.*, 726: 147—155.

Whiteford, P.C., 1976. Assessment of the audio-magneto-telluric method for geothermal resistivity surveying. In: *Proceedings 2nd U.N. Symposium on the Development and Use of Geothermal Resources*. U.S. Government Printing Office, Washington, D.C., 2: 1255—1261.

Wood, B., 1973. Geothermal power. In: H.C.H. Armstead (Editor), *Geothermal Energy*. UNESCO, Paris, pp. 109—121.

Yamada, E., 1976. Geological development of the Onikobe Caldera and its hydrothermal systems. In: *Proceedings 2nd U.N. Symposium on the Development and Use of Geothermal Resources*. U.S. Government Printing Office, Washington, D.C., 1: 665—672.
Yamasaki, T. and Hayashi, M., 1976. Geologic background of Otake and other geothermal areas in North-Central Kyushu, southwestern Japan. In: *Proceedings 2nd U.N. Symposium on the Development and Use of Geothermal Resources*. U.S. Government Printing Office, Washington, D.C., 1: 673—686.
Young, C.Y. and Ward, R.W., 1978. 2-D inversion of seismic attenuation observations in Coso hot springs. *KGRA Geotherm. Resour. Counc. Trans.*, 2: 747.
Young, C.Y. and Ward, R.W., 1980. Attenuation of teleseismic P-waves in The Geysers—Clear Lake region, California. *U.S. Geol. Surv., Prof. Paper* (in press).
Young, H.W. and Whitehead, R.L., 1975a. Geothermal investigations in Idaho, 2. An evaluation of thermal water in the Bruneau-Grand view area, southwest Idaho. *Idaho Dep. Water Resour., Water Inf. Bull.*, 30: 125 pp.
Young, H.W. and Whitehead, R.L., 1975b. Geothermal investigations in Idaho, 3. An evaluation of thermal water in the Weiser area, Idaho. *Idaho Dep. Water Resour., Water Inf. Bull.*, 30: 35 pp.

Zohdy, A.A., Anderson, L.A. and Muffler, L.J.P., 1973. Resistivity, self-potential and induced polarization surveys over a vapor-dominated geothermal system. *Geophysics*, 38: 1130—1144.

AUTHOR INDEX

SUBJECT INDEX